The Future of Marine Life in a Changing Ocean

The Fate of Marine Organisms and Processes under Climate Change and other Types of Human Perturbation

Series on the Science of Climate Change

ISSN: 2045-9726

Editor: Hans-F Graf *(University of Cambridge, UK)*

Published

Vol. 2   *The Future of Marine Life in a Changing Ocean: The Fate of Marine Organisms and Processes under Climate Change and other Types of Human Perturbation*
by M. Debora Iglesias-Rodriguez

Vol. 1   *Parameterization of Atmospheric Convection (In 2 Volumes)*
Volume 1: Theoretical Background and Formulation
Volume 2: Current Issues and New Theories
edited by Robert S. Plant and Jun-Ichi Yano

2
Series on the Science of Climate Change

The Future of Marine Life in a Changing Ocean

The Fate of Marine Organisms and Processes under Climate Change and other Types of Human Perturbation

M. Debora Iglesias-Rodriguez
University of California, Santa Barbara, USA

NEW JERSEY • LONDON • SINGAPORE • BEIJING • SHANGHAI • HONG KONG • TAIPEI • CHENNAI • TOKYO

Published by

World Scientific Publishing Europe Ltd.
57 Shelton Street, Covent Garden, London WC2H 9HE
Head office: 5 Toh Tuck Link, Singapore 596224
USA office: 27 Warren Street, Suite 401-402, Hackensack, NJ 07601

Library of Congress Cataloging-in-Publication Data
Names: Iglesias-Rodriguez, Debora (Maria Debora), author.
Title: The future of marine life in a changing ocean : the fate of marine organisms and processes
    under climate change and other types of human perturbation / by M. Debora Iglesias-Rodriguez
    (University of California, Santa Barbara, USA).
Description: New Jersey : World Scientific, [2019] | Series: Series on the science of
    climage change, 2045-9726 ; volume 2 | Includes bibliographical references.
Identifiers: LCCN 2019015572 | ISBN 9781786347428 (hardcover)
Subjects: LCSH: Marine ecology. | Climatic changes.
Classification: LCC QH541.5.S3 I35 2019 | DDC 577.7--dc23
LC record available at https://lccn.loc.gov/2019015572

British Library Cataloguing-in-Publication Data
A catalogue record for this book is available from the British Library.

Copyright © 2020 by World Scientific Publishing Europe Ltd.

All rights reserved. This book, or parts thereof, may not be reproduced in any form or by any means, electronic or mechanical, including photocopying, recording or any information storage and retrieval system now known or to be invented, without written permission from the publisher.

For photocopying of material in this volume, please pay a copying fee through the Copyright Clearance Center, Inc., 222 Rosewood Drive, Danvers, MA 01923, USA. In this case permission to photocopy is not required from the publisher.

For any available supplementary material, please visit
https://www.worldscientific.com/worldscibooks/10.1142/Q0220#t=suppl

Desk Editors: Aanand Jayaraman/Jennifer Brough/Shi Ying Koe

Typeset by Stallion Press
Email: enquiries@stallionpress.com

To Nico, my darling son with whom I share my love for nature.

Preface

The motivation for writing this book was to provide readers with accessible information about how humans are changing the ocean's chemistry, physics and biology, and to raise awareness about the urgent problems the oceans face today. For this purpose, I brought together the state of our knowledge on major disruptors of the marine environment caused by the large number of humans on Earth and their activities. Despite going through times that can be depressing for those who appreciate and care about the natural world, I, like many others involved in research dealing with anthropogenic impacts on the ocean, have been lucky to study phenomena such as ocean acidification and warming in real time. Indeed, we are witnessing fast changes in seawater chemistry that are causing alterations in the composition of ecosystems and their function. For example, within months, any rise in atmospheric carbon dioxide is seen in the surface ocean, which requires that marine organisms adapt fast. Similarly, rising temperatures are causing marine life to migrate or alter their physiology to cope not only with warming but also with a decrease in the solubility of oxygen that occurs with warming. And perhaps the most disturbing ocean problem that we face today is plastic pollution, which has reached every corner of Earth because plastics are extremely hard to remove from marine ecosystems and have harmful impacts on life.

One of my main concerns is the fact that apart from the immediate consequences, there are long-term effects that are going to have lasting impacts not only on ecosystems but also on human health. For example, our reliance on oil and plastic as commodities of everyday life are polluting our environment to the extent of affecting hormonal balance of marine animals and humans, and increasing our exposure to cancer-causing compounds. Despite humans being able to reach other planets and solve complex problems that require bringing together engineering and science, dealing with deterioration (and in some cases, destruction) of the marine environment has not been a priority. This book will be a useful guide to undergraduate and graduate students, as well as the well-informed reader, to learn about the causes and consequences of five major environmental stressors. Although some background knowledge is required, this book builds on concepts explained through the text or in footnotes. My first hope is that this book will be useful to any student with a science background interested in how humans are changing the ocean. Secondly, although, as a scientist, I try to remain the *Honest Broker*,[1] as a citizen, I hope this book will serve environmental activists and advocates in furthering their understanding of the major losses of ecosystem diversity and function as well as the ocean's beauty and esthetic values if we do not take serious and global action at several levels of organization in society from individual to corporate.

---

[1] *The Honest Broker* by Roger A. Pielke, Jr, Cambridge University Press, presents the options of scientific decision-making by going beyond the traditional roles of the "pure scientist" or the "issue advocate".

About the Author

 **M. Debora Iglesias-Rodriguez** is a Biological Oceanographer interested in how human activities affect marine life and ocean processes. She was born in Vigo (Galicia, Spain) and spent summers at Playa de Patos, where she developed a love for all things marine. She holds a BSc in Biology, an MSc in Marine Microbiology (University of Santiago de Compostela, Spain), and a PhD in Carbon Physiology (Swansea University, UK). She held academic positions at the Universities of Bristol and Nottingham (UK) and the National Oceanography Centre (University of Southampton, UK). In 2013, Débora moved to the University of California Santa Barbara, where she leads a lab exploring how human activities impact marine life. Her research topics include impacts of ocean acidification, warming, and oil pollution on microbial diversity and function, focusing on how climate change impacts marine calcification, one of the most important ways by which Earth locks carbon away from the atmosphere on planetary scales. She has advised the UK Government on the effects of climate change on marine ecosystems and has been involved in the writing of white papers on ocean acidification. Her work on the effects of ocean acidification on coccolithophore calcification has featured in leading national newspapers worldwide. She loves spending time in Galicia, Hawaii and California, where she lives.

Acknowledgments

I thank Aaron Bagnell, Frauke Bagusche, Ilana Berman-Frank, Dylan Catlett, Steve Comeau, Anna James, Bethan Jones, Tanika Ladd, Ed Laws, Paul Matson, Cindy Pilskaln, Richard Lampitt, Alyson Santoro, and Zoe Welch for their insight and their useful comments on chapters and the conversations about the content of this book. I thank Spencer Barnitz, Ross Barrett, Sophie Cooper, Janice Jones, Terence Keel, Mario Lebrato, Pilar Rodriguez Seoane, Greg Scott, Azure Stewart, Richard Stoker, Laurent Thomas, and Fawn Weider for their friendship and for being supportive while I was writing this book.

    I would like to thank my mother, Rosa Seoane, for her love and personal guidance as well as discussions about phytoplankton that fed into her artwork and resulted in fruitful art–science collaborations and the cover of this book. Thanks to my 103 years old grand mother, Carmen Seoane Cadavid, who was a feisty natural born leader and a true environmentalist. Thanks to my lil sister Noemi Iglesias Rodriguez, whose wisdom and advice have been invaluable. Your research on child poverty makes me think about possible links with environmental advocacy and justice (perhaps a future book for us?). Many thanks to my dear brother Pablo Iglesias Rodriguez, also an academic, whose steps and advice I follow from across the pond. And lastly, special thanks to my niece Sofía, who possesses the childhood wisdom that is lost when we move into adulthood and who was an inspiration for writing this book. Thanks for the great conversations about life and for the wonderful visits to her school to speak about climate change.

Contents

Preface vii

About the Author ix

Acknowledgments xi

Chapter 1. Introduction — Fundamental Concepts and Misconceptions 1

Chapter 2. Ocean Acidification — Human Activities Increase Atmospheric CO_2, Which Diffuses in the Surface Ocean Decreasing Seawater pH 27

Chapter 3. Ocean Warming — Continuing Increases in Greenhouse Gases from Human Activities Are Warming Earth 53

Chapter 4. Ocean Deoxygenation — Warming and Coastal Pollution Are Causing the Oceans to Become Oxygen-Starved 93

Chapter 5. Plastic Pollution — Excessive Plastic Production and Consumption Is Filling Up the Oceans 115

Chapter 6.	Oil Pollution — The Release of Petroleum Hydrocarbon into the Ocean from Its Extraction, Transportation, Refining, Storage, and Use Is Harming Marine Life	149
Chapter 7.	Thoughts on the Effects of Climate Change on Food Security	177
Index		201

Chapter 1

Introduction
Fundamental Concepts and Misconceptions

"... it is our culture, our livelihood, our economy and, for many, the ocean is the mother of all things."

— by Dame Meg Taylor, speaking in her role as Pacific Ocean Commissioner, describing the ocean as central to Pacific lives (Taylor, 2015).

Overview

The oceans play a pivotal role in the preservation of life on Earth. They represent two-thirds of the Earth's surface as ecosystems for marine animals and plants, supply essential sources of food, economic activity and biodiversity, and are central to the global cycling of elements, for example, carbon and nitrogen. Indeed, marine ecosystems provide great services to humans, for example, they provide a significant source of food globally: up to 20% of protein in Southeast Asian countries, Western Coastal Africa and Western and Northern Europe (FAO, 2007). The oceans also absorb about one-third of the carbon emissions from fossil fuel combustion, which in part ameliorates global warming (IPCC, WMO and UN Environment Programme, 2000; IPCC, 2001).

The human population is forecasted to reach close to 10 billion by 2070 (UN DESA, 2017), which is $\sim$10 times larger than in 1800 (Bongaarts, 2009) and it is considered to be within estimates of the maximum carrying capacity of the planet (Cohen, 1995). This will result in major increased demands for energy (almost double), and food and water (over 50%) (Ferroukhi et al., 2015). An urgent question for science is whether Earth's productivity can meet the demands of a human population of that magnitude.

Since 2008, approximately 24 million people have been displaced by catastrophic weather disasters each year (Global Internal Displacement Database, 2018), and given that warming is likely to amplify the frequency and intensity of natural disasters (Webster et al., 2005), it is expected that this number will rise. With climate insecurity[1] comes human migration, and climate refugees and climate migrants, both seen as a failure to mitigate climate but also as a strategy to adapt to climate change, based on the ideas of climate vulnerability, exposure to risk and adaptive capacity (see McLeman and Smit, 2006; Felli, 2012). Interestingly, climate change has also been found to pose a threat to mental health in humans as it can cause physical injury, psychological trauma, infrastructure damage, and societal disruption in affected regions (Obradovich et al., 2018). Although the details of how humans will adapt to climate change are under debate, human adaptation will undoubtedly come at a significant cost.

A major problem that has persisted since the beginning of Industrialization is marine pollution, which perhaps has lost the attention it deserves as global warming has been at the center of public debate. Alarmingly, more than half of our oil pollution has originated from illegal activities that include the dumping of ballast water and oil residues as well as accidents (Golyshin et al., 2003). In the case of plastic pollution, although the global governance of plastic is slowly improving (e.g., some companies such as Toyota, Evian, Asda, Waitrose are reducing plastic waste to landfills; activism intensifies; towns, cities, and legislatures are banning some uses of plastics), and as research continues to lead the design of less environmentally harmful plastic substitutes, the amount of plastic flowing into the oceans is on track to double from 2010 to 2025 (Jambeck et al., 2015). Although not covered in this book, submarine groundwater discharge, defined as "direct groundwater outflow across the ocean–land interface into the ocean" (Church, 1996), and coastal groundwaters, which are often contaminated with sewage, fertilizers, pathogens, pesticides or

---

[1] A condition under which the effects of climate variability and/or change are represented as a threat to a group of affected individuals (Trombetta, 2008; Mason, 2014).

industrial waste, can be important pathways for coastal pollution affecting the balance of coastal ecosystems as well as human health (UNESCO, 2004).

This chapter introduces some fundamental concepts and ideas that will be discussed in detail in the subsequent chapters and sets the stage for exploring major climate change phenomena. There are many other climate-driven phenomena either not mentioned in this book or not covered in any detail, for example, sea-level rise, coastal eutrophication, and many forms of marine pollution other than plastic and oil pollution (e.g., pharmaceutical discharge, waste from mining operations, soil erosion in agricultural land). The following are some important facts to get started:

- The definition of climate change used in this book is the one applied by the Framework Convention on Climate Change: "the change of climate that is attributed directly or indirectly to human activity that alters the composition of the global atmosphere and that is in addition to natural climate variability observed over comparable time periods". IPCC uses a different definition of climate change: "any change in climate over time, whether due to natural variability or as a result of human activity" (IPCC, 2007).
- *Ocean acidification continues to be a concern.* While the carbon chemistry changes associated with an increase in carbon dioxide levels in the upper ocean may be beneficial to some photosynthetic organisms, ocean acidification can pose a threat to many organisms that make shells, plates, and other biomineralized structures made of calcium carbonate such as spicules (endoskeleton in sea urchins) as part of their life cycle. There is however a great deal of variability of responses, which has created controversy around the fate of many calcifying groups and the relative importance of ocean acidification and other phenomena driven by human activities.
- *The oceans continue to get warmer.* A study revealed that ocean warming is at the high end of previous estimates, with implications for policy-related assessment of Earth's response to climate change (Resplandy *et al.*, 2018), such as climate sensitivity to greenhouse gases like CO_2 (Forster, 2016) and the thermal component of sea-level rise (Church *et al.*, 2013).
- *Marine pollution is an urgent issue.* Pollutants, some of which known to cause cancer and disrupt the hormonal balance in animals, can persist in the oceans for decades through centuries or even millennia. For example, persistent organic pollutants and toxic plastic leachates that cause harm

to marine organisms continue to be on the rise. Plastic pollution has now reached epic proportions: no part of the ocean can escape plastic pollution, and encountering plastic in seafood is becoming unavoidable. Plastic in the ocean is projected to triple by 2025 (U.K. Government Office for Science, 2018).
- *Anthropogenic perturbation impacts ecosystem diversity and function.* Although it is not always straightforward to predict what marine species and in what way they will adapt, and the exact way in which marine ecosystems will change, we know that ecosystems will look different from today because of climate change and other types of anthropogenic perturbation. We already know that humans are impacting ocean health and the services the oceans provide, such as carbon sequestration and food security.

First Humans Wrecked Terrestrial Ecosystems and Then the Oceans

In the early days of writing this book, I tried to stay away from the doom and gloom of climate change and the state of our oceans, as I thought it would be likely to distance or disengage individuals from climate science and rather leave them feeling helpless and overwhelmed. However, as I progressed with my writing and my research on each topic, the unfortunate gloomy reality became clear: nowhere in the ocean is free of human impacts (Halpern *et al.*, 2008; 2015), and marine wilderness has been damaged by humans, with only 13.2% lasting across the oceans (Jones *et al.*, 2018).

When looking at the impact of humans on biodiversity and habitat destruction, terrestrial ecosystems are a great example of how persisting human activities have led to species and population declines and extinctions, and alterations in ecosystem function and human welfare. For example, 322 species of terrestrial vertebrates have become extinct since 1500, and populations of the remaining species show 25% average decline in abundance (Dirzo *et al.*, 2014). In addition to the effect on ecosystem structure, species loss can impact human health and food security. The best example from terrestrial ecosystems is pollinators, which are key to the existence of many food crops and are strongly declining in abundance and diversity globally (Potts *et al.*, 2010). Similar to terrestrial ecosystems, the anthropogenically driven defaunation, and shifts in ecosystem services and function of the oceans began with the start of industrialization. It has been suggested that habitat degradation is likely to intensify as a major driver of marine wildlife loss and that proactive intervention can prevent a marine defaunation disaster of the proportions observed on terrestrial ecosystems (McCauley *et al.*, 2015).

The relationship that humans have with the oceans is rapidly changing and we will see a continued rise in activity and resource exploitation as large portions of terrestrial ecosystems continue to be depleted or severely damaged. The exploitation of the oceans as a source of renewable energy will disrupt marine ecosystems even further. The inevitable rise in aquaculture, given the depletion of important fisheries, also comes at a cost in terms of environmental and human health impacts. For example, both low environmental concentrations of antibiotics and legal aquaculture doses of antibiotics have been found to impair physiological functions, nutritional metabolism, and compromise fish immune system (Limbu *et al.*, 2018). Additionally, consumption of fish fed with legal oxytetracycline doses is a known health risk in children (Limbu *et al.*, 2018). Another risk of aquaculture practices is the increase of risk of disinfection byproducts, such as trihalomethanes and haloacetic acids that appear to have potential toxicity and have carcinogenic and mutagenic properties (Deng *et al.*, 2014; Richardson *et al.*, 2007). Also, pharmaceutically active compounds are expected to rise in the future (Asif *et al.*, 2018) and as they can accumulate in fish, they pose serious health concerns for human health (1) due to consumption of contaminated seafood/fish and (2) because exposure to these compounds can induce antibiotic resistance in humans and animals (Love *et al.*, 2011; Le *et al.*, 2005). All these extremely important problems the oceans and humans face are not discussed in this book although they deserve just as much attention.

Reading an old article by Holdren and Ehrlich (1974), I realized many humans perhaps still have a romanticized and idealized perception of the oceans. One of three widespread misconceptions from when I was a child regarding the impact of a growing human population, environmental deterioration, and resource depletion is still dangerously present today. We have moved forward and no longer believe the first one — that the size and growth rate of the human population has little or no relationship to the rapidly escalating ecological problems facing humankind. The second is that environmental deterioration consists primarily of pollution, which is perceived as a local and reversible phenomenon of concern mainly for its obvious and immediate effects on human health. Today, we know that environmental perturbation has both local and planetary implications and we are aware of the global impacts humans have created, such as warming and ocean acidification, that have been altering the physical and chemical properties of seawater worldwide. The third misconception, which persists today, is that science and technology can make possible the continued growth in human society via the consumption of natural resources with little consequences to the environment. I would add a fourth misconception — that the ocean somehow possesses unlimited inertia and the ability to return to steady state or "regenerate" itself. These last two dangerous misconceptions can lead to inaction or delayed intervention while there is plentiful evidence indicating that urgent solutions are required to prevent expansion of the problems.

Biogeochemical Functions and Trends Under Changing Climate

Among the globally impacting microbial functions, two processes that have been investigated in relation to climate change are photosynthesis and respiration. Photosynthesis is the process of converting carbon dioxide (CO_2) and water (H_2O) into organic matter and oxygen in the presence of light. Respiration, which is the reverse process, releases energy when organic matter is oxidized to produce CO_2 and H_2O.

$$6CO_2 + 6H_2O \underset{\text{Respiration}}{\overset{\text{Photosynthesis}}{\rightleftarrows}} C_6H_{12}O_6 + 6O_2$$

Respiration occurs not only in animals that have aerobic metabolism but also in plants (phytoplankton and cyanobacteria as well as macroalgae). As the oceans are getting warmer, the solubility of gases, including CO_2 and O_2, decreases. Indeed, CO_2 and O_2 concentrations in seawater decrease with ocean warming, but because CO_2 is accumulating in the atmosphere, its levels in the upper ocean are in fact increasing because the CO_2 in the surface ocean is in equilibrium with the CO_2 in the atmosphere. This increase in CO_2, which incidentally causes a decline in pH (hence the term "ocean acidification" — see Chapter 2), is expected to favor some photosynthetic organisms (see Connell et al., 2013) but can also have detrimental (Semesi et al., 2009) or little (e.g., Hoppe et al., 2018) effects on productivity, the latter suggesting that at least some components of the ecosystem can buffer changes in CO_2 levels. Additionally, a rise in the $CO_2:O_2$ ratio in the ocean surface is likely to affect carbon physiology in many photosynthetic organisms, particularly those living in conditions of low oxygen at the surface. This can be the case in some eastern boundary upwelling[2] systems where conditions of extremely low oxygen and even anoxia (absence of oxygen) can occur in near-surface waters in addition to elevated CO_2 (Bograd et al., 2008). In animals, these hypoxic conditions require additional energy requirements in order for the animals to adapt, e.g., by migrating to cooler waters or altering physiology or behavior (see Chapter 4).

A biogeochemically important function that appears to be negatively affected by climate change is calcification (see Chapter 2), specifically under increasingly "acidified" oceans. However, there are many exceptions and outliers of the expected decreasing trend of calcification responses to acidification, thus hampering our ability to make predictions with a reasonable degree of certainty. Indeed, since the first reports showing that calcification could increase in coccolithophores (a type of phytoplankton that produces $CaCO_3$) (Iglesias-Rodriguez et al., 2008) and other calcifying organisms (Ries et al., 2009) in response to ocean acidification, any possible calcification trend has been observed — increase, decrease, non-uniform, neutral — in both lab manipulations and field observations using numerous future climate scenarios (e.g., Ries et al., 2009; Kroeker et al., 2010). Finally, combined effects of ocean acidification and other climate-driven phenomena (e.g., warming) can have very different outcomes compared to those using single stressors (see Rodolfo-Metalpa et al., 2011; Harvey et al., 2013) adding to the challenge of representing and predicting calcification trends.

Microbial Functional Groups Governing Elemental Cycles

A diverse microbial consortium in marine ecosystems is responsible for nearly half of global primary production (Arrigo, 2005). At the base of the marine food web are bacteria and microscopic plants, and animals that have unique biogeochemical functions. For example, heterotrophic bacteria transform organic matter into dissolved inorganic nutrients (e.g., phosphate, nitrate, carbon dioxide) that end up being used by other microorganisms.

---

[2]Upwelling is a natural wind-driven process by which water rises up to the sea surface bringing up colder and nutrient-rich water. Upwelling zones are typically highly productive as the nutrients brought up by these waters "fertilize" the ocean surface.

For example, phytoplanktonic cells utilize dissolved inorganic nutrients and light energy to produce organic matter that ends up as food for higher (zooplankton) or lower (bacteria) trophic levels.

Some of these microorganisms are responsible for biogeochemical processes of global importance (e.g., silicification, calcification, nitrogen fixation, sulphur release) and can be grouped based on their specific functions. The term "functional group" refers to those organisms that are not necessarily phylogenetically related but rather those that share an important ecological or biogeochemical trait (Cummins, 1974; Steneck, 2001; Blondel, 2003). This concept has been applied mainly to marine microbes (Figure 1), e.g., diatoms, coccolithophores, dinoflagellates, oil-degrading bacteria, N_2-fixing bacteria, all responsible for important biogeochemical processes.

Examples of biogeochemically important functional groups are diatoms and coccolithophores which are phytoplankton groups that produce

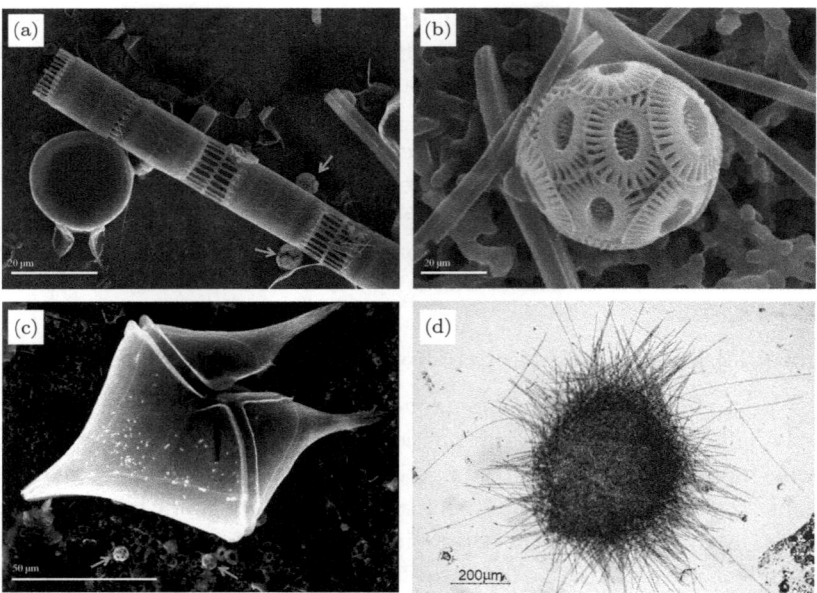

Figure 1. Examples of microbial functional groups. Scanning electron micrographs of (a) diatoms (silicifiers), (b) a coccolithophore (calcifier), and (c) a dinoflagellate, a group which comprises organisms specialized in different functions (e.g., bioluminescence, toxin production). (d) A light microscope image of the cyanobacterium *Trichodesmium*, an important nitrogen fixer that forms colonies called "puffs". Red arrows in images (a) and (c) point to coccolithophore cells. Images (a)–(c) were provided by Dr. Paul Matson and image (d) was provided by Dr. Ilana Berman-Frank.

biominerals of silica and calcium carbonate, respectively. Diatoms use silicic acid to make the intricate patterns of the external cell layer or frustule, which is composed almost entirely of silica. Coccolithophores use dissolved inorganic carbon not only for photosynthesis but also to precipitate beautiful ornate structures of calcium carbonate called coccoliths; dinoflagellates include groups of organisms with specialized functions (e.g., bioluminescence, toxin release, the production of armor containing cellulose); oil-degrading bacteria can break down hydrocarbons in areas of elevated hydrocarbon release, such as oil sips, and some can feed exclusively on hydrocarbons [for example, alkane oxidizers within *Desulfobacteraceae* degrade hydrocarbons in anaerobic conditions at marine seeps (Kleindienst *et al.*, 2014)]; diazotrophs fix atmospheric nitrogen (N_2) into dissolved inorganic nitrogen compounds (e.g., ammonium), which can be used by many microbes; and denitrifying bacteria use oxidized nitrogen compounds such as nitrate and nitrite as electron acceptors for energy *via* various stepwise reactions that result in the loss of N_2 to the atmosphere. The functional group concept can be expanded not just to microbes but also to multicellular organisms that share a particularly property. Accordingly, the silicifiers can include diatoms and also siliceous sponges and radiolarians; calcifiers can include corals, foraminifera and pteropods, as well as coccolithophores.

Functional Group Responses to Climate Change

Microbes are the great survivors to environmental change because of their huge genetic heterogeneity mostly due to fast division rates and their capacity to acquire novel genes via lateral gene transfer.[3] Also, given their unique life-cycle strategies (e.g., cysts, dormant phases), some microbes are able to survive harsh conditions and the environment itself selects for specialized functions. For example, elevated hydrocarbon concentrations in seawater select for oil-degrading bacteria (Kostka *et al.*, 2011) and hydrothermal vent environments select for specialized bacteria (Takami *et al.*, 1997; 1999; Teske *et al.*, 2000) such as chemosynthetic bacteria[4] (Stewart *et al.*, 2005). The fast adaptation of microbes gives them the

---

[3] A mechanism by which genes are transferred between taxonomically unrelated organisms rather than from parents to the offspring.
[4] Chemosynthetic bacteria oxidize reduced inorganic compounds (e.g., sulfide) for energy and fix CO_2 for biomass synthesis (i.e., chemoautotrophy). Alternatively, they use single carbon compounds as both energy and carbon source (e.g., methanotrophy).

potential to adapt to climate change and any environmental perturbation affecting their ecosystems.

A shift in organismal physiology (e.g., changes in the rate of primary production, calcification rates, nitrogen fixation, denitrification) as a result of climate change has a direct impact on ecosystem function and the cycling of elements (for example, carbon, nitrogen) between the atmosphere, the upper ocean and the ocean interior. Often, shifts in the magnitude or direction of these functions (e.g., an increase/decrease in nitrification/ denitrification, causing alterations in the fluxes of nitrogen between the ocean and the atmosphere) will affect ecosystems at several levels of organization, for example, primary production, and the population structure of primary producers and consumers. If the cause for the shift is driven by a globally occurring phenomenon such as warming or ocean acidification rather than a localized impact (e.g., coastal pollution), the repercussions could be manifested on planetary scales (e.g., global warming caused primarily by rising CO_2).

In addition to the changes that an observer can make on ecological timescales, the geological record reveals some resilience and adaptation to changes in the environment but also periods of extinction under extreme conditions of abrupt climate change (Foster *et al.*, 2018). However, what appears to be clear is that there is tremendous heterogeneity of responses to past climate change, for example, in rates of coral adaptation to warming and their geographic variation in bleaching (Pandolfi *et al.*, 2011). In the case of ocean acidification, past events are rather imperfect analogues (Gibbs *et al.*, 2010; Klaus *et al.*, 2011) for current ocean acidification scenarios (e.g., the rate of pH change originated from fossil fuel emissions is faster at present), and it is hard to tease apart the distinct roles of co-occurring environmental stressors from observations in the geological record and contrast them with those occurring in the present day (Kump *et al.*, 2009).

The Biological Carbon Pump and the Microbial Carbon Pump

The biological carbon pump (BCP) is the transfer of carbon (particulate and dissolved) from the surface ocean to the deep sea. This process starts in the sunlit upper ocean where primary producers fix dissolved inorganic carbon (DIC) (e.g., CO_2) into particulate organic carbon (POC) via photosynthesis. In addition to the formation of POC through photosynthesis, phytoplanktonic cells release dissolved organic matter (DOM) (including dissolved organic carbon) into seawater (Biddanda and Benner, 1997). Carbon (both particulate and dissolved) is transferred as both living and dead matter to the deep sea and the ocean sediment by gravity and by zooplankton migrating to depth. Because this process starts with the removal of CO_2 by primary producers at the ocean surface and its production at depth, and given that POC is utilized by bacteria and archaea (Aristegui et al., 2009), the BCP creates a CO_2 gradient with depth.

The efficiency of the BCP has been used as the fundamental measure of the ocean's capacity to store biologically fixed carbon. However, the conventional interpretation of the BCP has been revised recently because of its inefficiency in transferring carbon into the deep sea and the ocean floor given that the majority of carbon fixed into phytoplankton biomass each year is mineralized in the top few hundred meters of the ocean (Field et al., 1998; Martin et al., 1987). Only between 5% and 25% of primary production is exported from the euphotic zone to depth, and a mere ~0.3% reaches the deep sea and is buried in the sediment (De la Rocha and Passow, 2007; Zhang et al., 2018). So, where does the fixed carbon that is not mineralized go?

It appears that a portion of the carbon fixed by primary producers that is not mineralized can be stored for thousands of years as recalcitrant dissolved organic matter (RDOM[5]) as a result of what is referred to as the microbial carbon pump (MCP). Bacteria, archaea and viruses are responsible for the breakdown of most of the carbon that sinks to the deep sea (Suttle, 2007; Aristegui et al., 2009). In addition to phytoplankton contributing to DOM production, viral infections are responsible for the release of DOM via cell lysis (Suttle, 2007), and "sloppy feeding" by metazoan grazers can also release DOM. While POC is the focus of the BCP, the MCP processes centre around the production of DOM and specifically the pool of RDOM, which represents the majority of dissolved organic carbon and can remain at any depth and for long time periods (Bauer et al., 1992; Hansell, 2013; Zhang et al., 2018). Also, in contrast with the

---

[5]There are three types of DOM according to their biological availability (see Jiao et al., 2010): labile DOM (lDOM), semi-labile DOM (slDOM) and recalcitrant DOM (RDOM) (Eichinger et al., 2006; Calson and Ducklow, 1995). lDOM can be removed by heterotrophic microorganisms within hours or days; slDOM can persist for months to years and accounts for most of the DOM that is exported from the euphotic zone to the deep sea; and RDOM is the most persistent carbon pool, with the potential to remain in the ocean interior for thousands of years (see Jiao et al., 2010).

BCP, with primary producers being the main players, the MCP's main processes are governed by heterotrophic organisms and viruses. In addition to the MCP, microenvironmental conditions such as oxygen depletion near marine microbes could also be conducive to RDOM formation at any depth (Jiao et al., 2010). In summary, the dissolved component of carbon and thus the MCP must be considered in addition to the BCP when studying the ocean as a carbon reservoir (Jiao et al., 2010).

Carbon Sequestration

Although the term "carbon sequestration" has been used broadly to describe the uptake of carbon in the surface ocean or its export to depth without any temporal context, this term has a very specific meaning. Carbon sequestration is the amount of carbon transferred from the ocean surface to depths below the depth of the winter mixing (200–1000 m) that remains at those depths for a time period exceeding a 100 years (IPCC, 2007; Lampitt et al., 2008; Passow and Carlson, 2012). Indeed, the term "sequestration" refers to the long-term storage of carbon, a mechanism by which Earth locks carbon away from the atmosphere. According to this definition, losses by sinking from the euphotic zone are referred to as the "export flux", in contrast with the "sequestration flux" to deep water (Figure 2).

Carbon sequestration has been proposed as a way to tackle ocean problems, including counteracting the atmospheric and marine accumulation of CO_2 from human activities. Indeed, many geoengineering initiatives have focused on enhancing CO_2 removal at the ocean surface although a few have explored ways to promote carbon export to depth (Lampitt et al., 2008; Smetacek, 2018). For example, geoengineering approaches to offset the rise in CO_2 in the upper ocean by enhancing the biological and microbial carbon pumps include ocean fertilization experiments such as iron enrichment experiments (e.g., Martin et al., 1994; Boyd et al., 2000) and manipulating the environment to induce upwelling artificially or enhance it where upwelling occurs seasonally (Aure et al., 2007; Shepherd et al., 2007; Iglesias-Rodriguez et al., 2017; Pan et al., 2018).

Perhaps the best known fertilization experiments in the sunlit surface ocean are those adding iron to iron-depleted waters that are otherwise rich in macronutrients (e.g., nitrate, phosphate). These experiments have been typically conducted in the high-nutrient low-chlorophyll (HNLC) regions such as Antarctic waters, which become extremely productive when

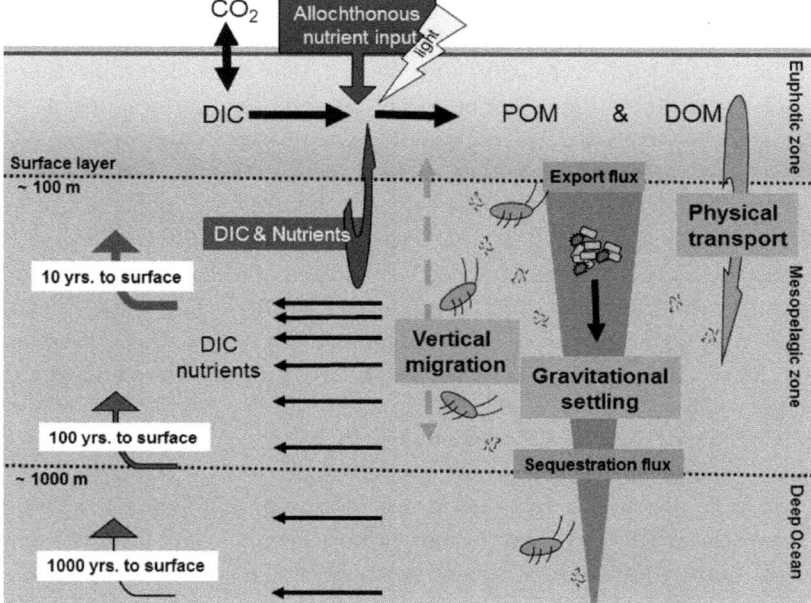

Figure 2. The biological carbon pump and the transfer of matter (particulate and dissolved) to the deep sea (by Passow and Carlson, 2012). The upper ocean, where light enables photosynthesis, phytoplankton produce organic matter that is transported to depth via different mechanisms (a) vertical migrating zooplankton (brown ovals with antennas), (b) gravitational settling of marine snow[6] (green clusters with black bacteria), or (c) physical transport of dissolved organic matter (DOM). The sinking flux of aggregates (represented by green cells and blue transparent exopolymer particles) decreases with depth as organic matter is mineralized (turned into its inorganic form). The average time for dissolved inorganic carbon (DIC) and nutrients to return back to the surface depends on the depth of mineralization. The sedimentation out of the surface water layer is termed export flux, whereas the sedimentation out of intermediate depths (the mesopelagic), between ~100 m and 1,000 m) is called sequestration flux. The carbon sequestration flux determines the storage of carbon within the ocean for at least a century (© Inter-Research 2012). See Passow and Carlson (2012) for details.

fertilized with iron. However, of the 13 major iron-fertilization experiments in the open ocean since 1990, few looked into carbon export to depth to

---

[6] Marine snow are particles >0.5 mm to 10s of cm in size, which may consist of aggregations of smaller organic and inorganic particles, including bacteria, phytoplankton, microzooplankton, zooplankton fecal pellets and feeding structures (e.g., larvacean houses), biominerals, terrestrially-derived lithogenic components, and detritus (Alldredge and Silver, 1988).

explore sequestration, which is in fact the key process to test as it governs carbon removal in the upper ocean in the long term.

The interest in artificially upwelling waters from depth to the sea surface stems from their potential to fertilize the surface ocean with nutrients characteristically more abundant in deep cold waters. Some criticisms of artificial upwelling include potential loss of CO_2 to the atmosphere as deep cold waters are typically not only rich in macronutrients and micronutrients (e.g., iron), but they are also rich in CO_2 because the solubility of CO_2 increases with decreasing temperature. As cold CO_2-rich waters rise towards the surface, a proportion of the CO_2 could potentially be lost to the atmosphere before it can be used in photosynthesis (Shepherd *et al.*, 2007). Also, given that deep waters are typically low in oxygen due to high respiration rates at depth, upwelling of these waters can result in hypoxia (low oxygen levels) and potentially anoxia (lack of oxygen at the surface) (Keller *et al.*, 2014; Pan *et al.*, 2016).

Overall, although both ocean iron fertilization and artificial upwelling (among other methods) have the potential to remove CO_2 from the ocean surface, both approaches appear to be ineffective in reducing warming on planetary scales and with substantial potential side effects (see Keller *et al.*, 2014).

The Conspicuous and Invisible Realities of Plastic and Oil Pollution

Since fossil fuels (coal, oil, gas) and plastic became indispensable to the functioning of many aspects of human society, oil and plastic pollution have become major environmental problems in the oceans. Despite not being the major source of oil pollution, what tends to attract public attention are major oil spills that cause environmental disasters with noticeable and disturbing consequences for marine life. Given that oil reserves are finite and non-renewable, and the world demand for oil is unsustainable, energy security is a concern and an important consideration in the development of future fuel. Clean alternatives include nuclear energy and hydrogen gas, but both have important safety risks (Crowl and Jo, 2007; Christodouleas *et al.*, 2011). Similarly, the accumulation of plastic in the open ocean and coastal environments, the fatalities caused by plastic debris in marine animals (e.g., suffocation, internal and external injuries, starvation caused by plastic ingestion), and the increasing finding of plastic particles in marine organisms, causing a concern for food security, are at the centre of public debate.

Although these are major problems that require urgent solutions, the invisible effect of major pollutants such as oil and plastic can cause major disruptions to marine animal and human health and can remain in the ecosystem for a long time. For example, polycyclic aromatic hydrocarbons (PAHs) and other components of oil as well as plastic are endocrine disruptors (Rochman et al., 2014; Zhang et al., 2016), and the small particles of plastic accumulate in animals after ingestion of organisms containing micro and nanoplastics (see Ziccardi et al., 2016). Dioxin-like PAHs or polychlorinated biphenyl (PCB) are toxic, and environmentally hazardous and carcinogenic even in small amounts (Chen et al., 2019). Effects of such dioxin-like chemicals include hepatotoxicity, certain types of cancer, immunotoxicity, and reproductive toxicity (Eichbaum et al., 2014). The perverse nature of this pollution is that its effects (e.g., poor health, death) are often not seen immediately as continued exposure leads to bioaccumulation within organisms and biomagnification in the food chain reaching humans.

The future of the oceans as a source of food

Hunger and food insecurity continue to be the most distressing problems that the poorest and most disadvantaged people in the world face (FAO, IFAD and WFP, 2015; NRC, 2006; Tacon, 2001; Tacon and Metian, 2018). For example, between 2014 and 2016, more than 795 million people had chronically inadequate levels of dietary energy intake (i.e., caloric intake) (FAO, IFAD and WFP, 2015). Moreover, deficiencies in micronutrients (e.g., iron, vitamins) referred to as the "hidden hunger", is even more widespread (Tacon and Metian, 2018). Paradoxically, it is estimated that more than 1.5 billion people are overweight and 0.5 billion are obese, increasing their health risks including cardiovascular problems and high blood sugar as well as other diet-related, non-communicable diseases (Sundaram et al., 2015).

The oceans provide a significant percentage of humans' dietary protein intake. In 2013, fish accounted for 20% of the global population's intake of animal protein and 6.7% of all protein consumed (FAO, 2016). Within Asia and many lower income countries within the African region, fish and fishery products represent the major source of dietary animal protein consumed (Tacon and Metian, 2018). However, the continuous expansion of some fisheries and aquaculture operations means that the risks of overfishing and the unsustainable use of natural resources are on the rise. In addition to overfishing, pollution, environmental degradation, impacts from growing coastal populations, climate change, diseases and natural and human-induced disasters add to the threats to fishers' livelihoods (FAO, 2016).

What is less discussed in the context of marine food security is the alarming decline (although not yet quantified) in food quality and safety (in addition to food quantity) because the oceans are becoming extremely polluted. Following the definition of food security by the Food and Agricultural Organization (FAO), one element of this term is that food must be safe for consumption (FAO, 1996).[7] Worryingly, the oceans are reaching a critical point concerning some pollutants because the oceans have been used as the dumping and waste ground for all kinds of activities from households to mining. An example is microplastics, that have been at the center of public debate because of their accumulation in seawater and plastics have been detected in commercial seafood including fish and bivalves both in wild and farmed species (Antão Barboza et al., 2018).

Dioxins, which are one of the dangerous chemicals listed in the Stockholm Convention on Persistent Organic Pollutants (POP), are another type of marine pollutant that has been linked to cancer and changes in immune and endocrine function (International Agency for Research on Cancer, 1997; FAO/WHO, 2010; Booth et al., 2013). Dioxins also belong to the so-called "dirty dozen" and they are of great concern because of their high toxicity (ENB, 2005). The oceans are a recipient of dioxins as 40% of airborne dioxin emissions are deposited to marine environments (FAO/WHO, 2010), resulting in dioxin accumulation in fish and seafood (Agency for Toxic Substances and Disease Registry, 1999).

Mercury is another extremely harmful pollutant that has reached the ocean through human activities, for example, from illegal or artisanal gold mines in Asia that release mercury, which is subsequently taken up and transported by atmospheric and oceanic currents, and transformed into methylmercury (Probyn, 2018). Interestingly, the flow of mercury appears to be gender- and ethnicity-biased as women workers in Indonesia are among those primarily affected as well as women in the northern hemisphere, which are the recipients of methylmercury toxicity from fish (Probyn, 2018). In adults, continuous intake of methylmercury has the potential to lead to neurological damage, kidney malfunction, and

---

[7] A condition when all people, at all times, have physical and economic access to sufficient, safe and nutritious food to meet their dietary needs and food preferences for an active healthy life (FAO, 1996).

reproductive impairment through damaging of sperm or the fetus (Agency for Toxic Substances and Disease Registry, 1999; Bostic *et al.*, 2018).

In reaction to the poor state of the oceans, the idea of the "Blue Economy" or "Oceans Economy"[8] arose as a model to incorporate ocean values and services into economic forecasting and decision-making. A motivation for the Blue Economy paradigm was to create a sustainable development framework for developing countries addressing equity in access to, development of and the sharing of benefits from marine resources; offering scope for reinvestment in human development and the alleviation of crippling national debt burdens (United Nations Conference on Sustainable Development, 2012). Governments and regional agencies of the Pacific Islands are already strengthening their commitment to sustainable oceans management through proactive policies and programs (Smith-Godfrey, 2016). Although the Blue Economy concept is increasingly being proposed, clarity on definitions and implementation steps remain vague (Smith-Godfrey, 2016).

Key Points

- Climate change and the rapidly growing human population, expected to reach close to 10 billion by the middle of the century, are increasing demands for energy and food of marine origin, placing a burden on the oceans.
- In addition to an increased demand for food being a threat to the capacity of the oceans to fulfill the needs of the human population, human activities are releasing toxic pollutants that are affecting the health of marine biota as well representing a risk to human health.
- The ocean has a limited "inertia" in dealing with human perturbation and does not possess the unlimited capacity to return to steady state or "regenerate" itself. Science and technology can only ameliorate some human impacts on the ocean but cannot make up for the impacts caused by the continued growth in the human population and non-sustainable activities.

---

[8] Defined as "the sustainable industrialization of the oceans to the benefit of all" (United Nations Conference on Sustainable Development, 2012) and later as "a sustainable ocean economy (which) emerges when economic activity is in balance with the long-term capacity of ocean ecosystems to support this activity and remain resilient and healthy" (Goddard, 2015).

- A rise in CO_2 has the potential to enhance photosynthetic carbon fixation and it will detrimentally affect some (but not all) calcium carbonate-making organisms. A decline in O_2 will require marine animals to adapt and spend energy to cope with low-oxygen conditions. In extreme conditions (e.g., anoxia), individuals and populations can be severely affected and die.
- The increase in atmospheric CO_2 will cause a rise in the CO_2 in the upper ocean and an increase in the CO_2:O_2 ratio, which is likely to affect carbon physiology in photosynthetic organisms, particularly those living in conditions of low oxygen.
- The invisible pollution in the oceans is the most threatening pollution to the environment and to humans because many pollutants do not break down or are difficult to remove from seawater. Additionally, the effects of chemical pollution are often not seen immediately but after long-term exposure, making it difficult sometimes to assess the true danger of exposure.

References

Agency for Toxic Substances and Disease Registry (1999) Toxicological profile for mercury. US Department of Health and Human Services. http://www.atsdr.cdc.gov/toxprofiles/tp46.pdf.

Alldredge AL, Silver MW (1988) Characteristics, dynamics and significance of marine snow. *Progress in Oceanography* **20**, 41–82.

Antão Barboza LG, Vethaak AD, Lavorante BLBO, Lundebye A-K, Guilhermino L (2018) Marine microplastic debris: an emerging issue for food security, food safety and human health. *Marine Pollution Bulletin* **133**, 336–348.

Aristegui J, Gasol JM, Duarte CM, Herndl GJ (2009) Microbial oceanography of the dark ocean's pelagic realm. *Limnol. Oceanogr.* **54**, 1501–1529.

Arrigo KR (2005) Marine microorganisms and global nutrient cycles. *Nature* **437**, 349–355.

Asif MB, Hai FI, Price WE, Nghiem LD (2018) Impact of pharmaceutically active compounds in marine environment on aquaculture. In: Hai F., Visvanathan C., Boopathy R. (eds.), *Sustainable Aquaculture. Applied Environmental Science and Engineering for a Sustainable Future*. Springer.

Aure J, Strand Ø, Erga SR, Strohmeier T (2007) Primary production enhancement by artificial upwelling in a western Norwegian fjord. *Marine Ecology Progress Series* **352**, 39–52.

Bauer JE, Williams PM, Druffel ERM (1992) ^{14}C activity of dissolved organic carbon fractions in the north-central Pacific and Sargasso Sea. *Nature* **357**, 667–670.

Biddanda B, Benner R (1997) Carbon, nitrogen, and carbohydrate fluxes during the production of particulate and dissolved organic matter by marine phytoplankton. *Limnology and Oceanography* **42**, 506–518.

Blondel J (2003) Guilds or functional groups: Does it matter? *Oikos* **100**, 223–231.

Bograd SJ, Castro CG, Di Lorenzo E, Palacios DM, Bailey H, Gilly W, Chavez FP (2008) Oxygen declines and the shoaling of the hypoxic boundary in the California current. *Geophysical Research Letters* **35**, L12607, doi:10.1029/2008GL034185.

Bongaarts J (2009) Human population growth and the demographic transition. *Philosophical Transactions of the Royal Society B* **364**, 2985–2990.

Booth S, Hui J, Alojado Z, Lam V, Cheung W, Zeller D, Steyn D, Pauly D (2013) Global deposition of airborne dioxin. *Marine Pollution Bulletin* **75**, 182–186.

Bostic SM, Sobal S, Bisogni CA (2018) Social representations of fish and seafood among midlife rural adults: Benefits, risks, and involvement. *Food Policy* **76**, 99–108.

Boyd PW, Watson AJ, Laws CS, Abraham ER *et al.* (2000) Phytoplankton bloom upon mesoscale iron fertilisation of polar Southern Ocean waters. *Nature*, **407** 695–702.

Carlson CA, Ducklow HW (1995) Dissolved organic carbon in the upper ocean of the central equatorial Pacific Ocean, 1992: daily and finescale vertical variation. *Deep-Sea Research II* **42**, 639–656.

Chen Q, Zhang H, Allgeier A, Zhou Q, Ouellet JD, Crawford SE, Luo Y, Yang Y, Shi H, Hollert H (2019) Marine microplastics bound dioxin-like chemicals: Model explanation and risk assessment. *Journal of Hazardous Materials* **364**, 82–90.

Christodouleas JP, Forrest RD, Ainsley CG, Tochner Z, Hahn SM, Glatstein E (2011) Short-term and long-term health risks of nuclear-power-plant accidents. *New England Journal of Medicine* **364**, 2334–2341.

Church TM (1996) An underground route for the water cycle. *Nature* **380**, 579–580.

Church JA, Clark PU, Cazenave A, Gregory JM, Jevrejeva S, Levermann A, Merrifield MA, Milne GA, Nerem RS, Nunn PD, Payne AJ, Pfeffer WT, Stammer D, Unnikrishnan AS (2013) Sea level change. In: Stocker T. F., Qin D., Plattner G.-K., Tignor M., Allen S. K., Boschung J., Nauels A., Xia Y., Bex V., and Midgley P. M. (eds.), *Climate Change 2013: The Physical Science Basis. Contribution of Working Group I to the Fifth Assessment Report of the Intergovernmental Panel on Climate Change.* Cambridge University Press, Cambridge, United Kingdom and New York, NY, USA.

Cohen JE (1995) How Many People Can the Earth Support? Norton, NYREV Inc.

Connell SD, Kroeker KJ, Fabricius KE, Kline DI, Russell BD (2013) The other ocean acidification problem: CO_2 as a resource among competitors for ecosystem dominance. *Phil Trans R Soc B* **368**, 20120442. http://dx.doi.org/10.1098/rstb.2012.0442.

Crowl DA, Jo Y-D (2007) The hazards and risks of hydrogen. *Journal of Loss Prevention in the Process Industries* **20**, 158–164.

Cummins KW (1974) Structure and function of stream ecosystems. *BioScience* **24**, 631–641.

De La Rocha CL, Passow U (2007) Factors influencing the sinking of POC and the efficiency of the biological carbon pump. *Deep-Sea Research II* **54**, 639–658.

Deng Y, Zhang Y, Zhang R, Wu B, Ding L, Xu K, Ren H (2014) Mice *in vivo* toxicity studies for monohaloacetamides emerging disinfection byproducts based on metabolomic methods. *Environmental Science and Technology* **48**, 8212–8219.

Dirzo R, Young HS, Galetti M, Ceballos G, Isaac NJB, Collen B (2014) Defaunation in the anthropocene. *Science* **345**, 401–406.

Eichbaum K, Brinkmann M, Buchinger S, Reifferscheid G, Hecker M, Giesy JP, Engwall M, van Bavel B, Hollert H (2014) *In vitro* bioassays for detecting dioxin-like activity — application potentials and limits of detection, a review, *Science of the Total Environment* **487**, 37–48.

Eichinger M, Poggiale JC, Van Wambeke F, Lefèvre D, Sempéré R (2006) Modelling DOC assimilation and bacterial growth efficiency in biodegradation experiments: a case study in the Northeast Atlantic Ocean. *Aquatic Microbial Ecology* **43**, 139–151.

ENB. Summary of the First Conference of the Parties to The Stockholm Convention. Earth Negotiations Bulletin; 2005, Vol. 15(117). http://enb.iisd.org/vol15/enb15117e.html.

FAO (1996) Rome declaration on world food security and World Food Summit plan of action. *World Food Summit*, Rome, 13–17 November 1996.

FAO (2007) The state of world fisheries and aquaculture 2006. *Food and Agriculture Organisation of the United Nations*, Rome, 180.

FAO/WHO (2010) Joint FAO/WHO Expert Consultation on the Risks and Benefits of Fish Consumption. http://www.fao.org/docrep/014/ba0136e/ba0136e00.pdf.

FAO, IFAD and WFP (Food and Agriculture Organization/International Fund for Agricultural Development/the World Food Programme) (2015) The state of food insecurity in the world 2015. Meeting the 2015 international hunger targets: Taking stock of uneven progress. Rome: FAO.

FAO (2016) The State of World Fisheries and Aquaculture 2016 (SOFIA): Contributing to food security and nutrition for all. Rome: Food and Agriculture Organization, 2016, pp. 200.

Felli R (2012) Managing climate insecurity by ensuring continuous capital accumulation: 'climate refugees' and 'climate migrants'. *New Political Economy* **18**, 337–363.

Ferroukhi R, Khalid A, Lopez-Peña A, Renner M (2015) Renewable energy and jobs: Annual review 2015. *International Renewable Energy Agency (IREA)*.

Field CB, Behrenfeld MJ, Randerson JT, Falkowski P (1998) Primary production of the biosphere: Integrating terrestrial and oceanic components. *Science* **281**, 237–240.

Forster PM (2016) Inference of climate sensitivity from analysis of Earth's energy budget. *Annual Review of Earth and Planetary Science* **44**, 85–106.

Foster GL, Hull P, Lunt DJ, Zachos JC (2018) Placing our current 'hyperthermal' in the context of rapid climate change in our geological past. *Philosophical*

Transactions of the Royal Society A **376**, 20170086. http://dx.doi.org/10.1098/rsta.2017.0086.

Gibbs SJ, Stoll HM, Bown PR, Bralower TJ (2010) Ocean acidification and surface water carbonate production across the Paleocene–Eocene thermal maximum. *Earth and Planetary Science Letters* **295**, 583–592.

Global Internal Displacement Database (2018) Internal Displacement Monitoring Centre, Geneva, Switzerland.

Goddard C (2015) The Blue Economy: Growth, opportunity and a sustainable ocean economy, s.l.: Economist Intelligence Unit. http://www.greengrowthknowledge.org/resource/blue-economy-growth-opportunity-and-sustainable-ocean-economy.

Golyshin PN, Martins Dos Santos VA, Kaiser O, Ferrer M, Sabirova YS, Lunsdorf H (2003) Genome sequence completed of *Alcanivorax borkumensis*, a hydrocarbon-degrading bacterium that plays a global role in oil removal from marine systems. *Journal of Biotechnology* **106**, 215–220.

Halpern BS, Walbridge S, Selkoe KA, Kappel CV, Micheli F, D'Agrosa C, Bruno JF, Casey KS, Ebert C, Fox HE, Fujita R, Heinemann D, Lenihan HS, Madin EMP, Perry MT, Selig ER, Spalding M, Steneck R, Watson R (2008) A global map of human impact on marine ecosystems. *Science* **319**, 948–952.

Halpern BS, Frazier M, Potapenko J, Casey KS, Koenig K, Longo C, Lowndes JS, Rockwood RC, Selig ER, Selkoe KA, Walbridge S (2015) Spatial and temporal changes in cumulative human impacts on the world's ocean. *Nature Communications* **6**, 7615.

Hansell DA (2013) Recalcitrant dissolved organic carbon fractions. *Annual Reviews of Marine Science* **5**, 421–445.

Harvey BP, Gwynn-Jones D, Moore PJ (2013) Meta-analysis reveals complex marine biological responses to the interactive effects of ocean acidification and warming. *Ecology and Evolution* **3**, 1016–1030.

Holdren JP, Ehrlich PR (1974) Human population and the global environment: Population growth, rising per capita material consumption, and disruptive technologies have made civilization a global ecological force. *American Scientist* **62**, 282–292.

Hoppe CJM, Wolf KKE, Schuback N, Tortell PD, Rost B (2018) Compensation of ocean acidification effects in Arctic phytoplankton assemblages. *Nature Climate Change* **8**, 529–533.

Iglesias-Rodriguez MD, Halloran PR, Rickaby REM, Hall IR, Colmenero-Hidalgo E, Gittins JR, Green DRH, Tyrell T, Gibbs SJ, von Dassow P, Rehm E, Armbrust EV, Boessenkool KP (2008) Phytoplankton calcification in a high-CO_2 world. *Science* **320**, 336–340.

Iglesias-Rodriguez MD, Jones BM, Blanco-Ameijeiras S, Greaves M, Huete-Ortega M, Lebrato M (2017) Physiological responses of coccolithophores to abrupt exposure of naturally low pH deep seawater. *PLoS ONE* **12**, e0181713.

Intergovernmental Panel on Climate Change, World Meteorological Organization & United Nations Environment Programme (2000) Land use, land-use change, and forestry. In: Watson R. T., Noble I. R., Bolin B., Ravindranath

N. H., Verardo D. J., Dokken D. J. (eds.), Cambridge University Press, Cambridge, UK.

Intergovernmental Panel on Climate Change (2001) Climate Change 2001: The Scientific Basis (Cambridge University Press, Cambridge, UK).

International Agency for Research on Cancer (1997) Polychlorinated dibenzo-p-dioxins and polychlorinated dibenzofurans: Summary of data reported and evaluation. *IARC Monographs on the Evaluation of Carcinogenic Risks to Humans* **69**, 33.

IPCC (2007) Climate Change 2007. Fourth Assessment Report to the Intergovernmental Panel on Climate Change (AR4). Cambridge University Press, Cambridge, UK.

Jambeck JR, Geyer R, Wilcox C, Siegler TR, Perryman M, Andrady A, Narayan R, Law KL (2015) Plastic waste inputs from land into the ocean. *Science* **347**, 768–771.

Jiao N, Herndl GJ, Hansell DA, Benner R, Kattner G, Wilhelm SW, Kirchman DL, Weinbauer MG, Luo T, Chen F, Azam F (2010) Microbial production of recalcitrant dissolved organic matter: Long-term carbon storage in the global ocean. *Nature Reviews* **8**, 593–599.

Jones KR, Klein CJ, Halpern BS, Venter O, Grantham H, Kuempel CD, Shumway N, Friedlander AM, Possingham HP, Watson JEM (2018) The location and protection status of Earth's diminishing marine wilderness. *Current Biology* **28**, 2506–2512.

Keller DP, Feng EY, Oschlies A (2014) Potential climate engineering effectiveness and side effects during a high carbon dioxide-emission scenario. *Nature Communications* **5**, 3304.

Klaus JS, Lutz BP, McNeill DF, Budd AF, Johnson KG, Ishman SE (2011) Rise and fall of Pliocene free-living corals in the Caribbean. *Geology* **39**, 375–378.

Kleindienst S, Herbst FA, Stagars M, von Netzer F, von Bergen M, Seifert J, Peplies J, Amann R, Musat F, Lueders T, Knittel K (2014) Diverse sulfate-reducing bacteria of the *Desulfosarcina/Desulfococcus* clade are the key alkane degraders at marine seeps. *The ISME Journal* **8**, 2029–2044.

Kostka JE, Prakash O, Overholt WA, Green S, Freyer G, Canion A, Delgardio, Norton N, Hazen TC, Huettel M (2011) Hydrocarbon-degrading bacteria and the bacterial community response in Gulf of Mexico beach sands impacted by the Deepwater Horizon oil spill. *Application in Environmental Microbiology* **77**, 7962–7974.

Kroeker KJ, Kordas RL, Crim RN, Singh GG (2010) Meta-analysis reveals negative yet variable effect of ocean acidification on marine organisms. *Ecology Letters* **13**, 1419–1434.

Kump LR, Bralower TJ, Ridgwell A (2009) Ocean acidification in deep time. Special issue on the future of ocean biogeochemistry in a high-CO_2 world. *Oceanography* **22**, 94–107.

Lampitt RS, Achterberg EP, Anderson TR, Hughes JA, Iglesias-Rodriguez MD, Kelly-Gerreyn BA, Lucas M, Popova E, Sanders R, Shepherd JG, Smythe-wright D, Yool A (2008) Ocean fertilization: A potential means of geoengineering? *Philosophical Transactions of the Royal Society A* **366**, 3919–3945.

Le TX, Munekage Y, Kato S-I (2005) Antibiotic resistance in bacteria from shrimp farming in mangrove areas. *Science of the Total Environment* **349**, 95–105.

Limbu SM, Zhou L, Sun S-X, Zhang M-L, Du Z-Y (2018) Chronic exposure to low environmental concentrations and legal aquaculture doses of antibiotics cause systemic adverse effects in Nile tilapia and provoke differential human health risk. *Environmental International* **115**, 205–219.

Love DC, Rodman S, Neff RA, Nachman KE (2011) Veterinary drug residues in seafood inspected by the European Union, United States, Canada, and Japan from 2000 to 2009. *Environmental Science and Technology* **45**, 7232–7240.

Martin JH, Knauer GA, Karl DM, Broenkow WW (1987) VERTEX: Carbon cycling in the northeast Pacific. *Deep-Sea Research* **43**, 267–285.

Martin JH, Coale KH, Johnson KS, Fitzwater SE, et al. (1994) Testing the iron hypothesis in ecosystems of the Equatorial Pacific Ocean. *Nature* **371**, 123–129.

Mason M (2014) Climate insecurity in (post)conflict areas: the biopolitics of United Nations vulnerability assessments. *Geopolitics* **19**, 806–828.

McCauley DJ, Pinsky ML, Palumbi SR, Estes JA, Joyce FH, Warner RR (2015) Marine defaunation: Animal loss in the global ocean. *Science* **347**, 1255641.

McLeman R, Smit B (2006) Migration as an adaptation to climate change. *Climatic Change* **76**, 31–53.

NRC (National Research Council) (2006) Food Insecurity and Hunger in the United States: An Assessment of the Measure. Panel to Review the U.S. Department of Agriculture's Measurement of Food Insecurity and Hunger. In: Gooloo S. Wunderlich and Janet L. Norwood (eds.), *Committee on National Statistics, Division of Behavioral and Social Sciences and Education*. Washington, DC: The National Academies Press.

Obradovich N, Migliorini R, Paulus MP, Rahwan I (2018) Empirical evidence of mental health risks posed by climate change. *Proceedings of the National Academy of Sciences* **115**, 10953–10958.

Pan YW, Fan W, Zhang DH, Chen JW, Huang HC, Liu SX, Jiang ZP, Di YN, Tong MM, Chen Y (2016) Research progress in artificial upwelling and its potential environmental effects. *Science China, Earth Sciences* **59**, 236–248.

Pan Y, You L, Li Y, Fan W, Chen C-TA, Wang BJ, Chen Y (2018) Achieving highly efficient atmospheric CO_2 uptake by artificial upwelling. *Sustainability* **10**, 664; doi:10.3390/su10030664.

Pandolfi JM, Connolly SR, Marshall DJ, Cohen AL (2011) Projecting coral reef futures under global warming and ocean acidification. *Science* **333**, 418–422.

Passow U, Carlson CA (2012) The biological pump in a high CO_2 world. *Marine Ecology Progress Series* **470**, 249–271.

Potts SG, Biesmeijer JC, Kremen C, Neumann P (2010) Global pollinator declines: trends, impacts and drivers. *Trends in Ecology and Evolution* **25**, 345–353.

Probyn E (2018) The ocean returns: Mapping a mercurial anthropocean. *Social Science Information* **57**, 386–402.

Resplandy L, Keeling RF, Eddebbar Y, Brooks MK, Wang R, Bopp L, Long MC, Dunne JP, KoeveW, Oschlies A (2018) Quantification of ocean heat uptake from changes in atmospheric O_2 and CO_2 composition. *Nature* **563**, 105–108.

Richardson SD, Plewa MJ, Wagner ED, Schoeny R, Demarini DM (2007) Occurrence, genotoxicity, and carcinogenicity of regulated and emerging disinfection by-products in drinking water: A review and roadmap for research. *Mutation Research/Fundamental and Molecular Mechanisms of Mutagenesis* **636**, 178–242.

Ries JB, Cohen AL, McCorkle DC (2009) Marine calcifiers exhibit mixed responses to CO_2-induced ocean acidification. *Geology* **37**, 1131–1134.

Rochman CM, Kurobe T, Flores I, The SJ (2014) Early warning signs of endocrine disruption in adult fish from the ingestion of polyethylene with and without sorbed chemical pollutants from the marine environment. *Science of the Total Environment* **493**, 656–661.

Rodolfo-Metalpa R, Houlbréque F, Tambutté E, Boisson F, Baggini C, Patti FP, Jeffree R, Fine M, Foggo A, Gattuso J-P, Hall-Spencer JM (2011) Coral and mollusc resistance to ocean acidification adversely affected by warming. *Nature Climate Change* **1**, 308–312.

Semesi IS, Kangwe J, Björk M (2009) Alterations in seawater pH and CO_2 affect calcification and photosynthesis in the tropical coralline alga, *Hydrolithon* sp. (Rhodophyta). *Estuarine, Coastal and Shelf Science* **84**, 337–341.

Shepherd JG, Iglesias-Rodriguez MD, Yool A (2007) Geo-engineering might cause, not cure, problems. *Nature* **449**, 781.

Smetacek V (2018) Seeing is believing: Diatoms and the ocean carbon cycle revisited. *Protist* **169**, 791–802.

Smith-Godfrey S (2016) Defining the Blue economy, maritime affairs. *Journal of the National Maritime Foundation of India* **12**, 58–64.

Steneck RS (2001) Functional groups. In: Levin, SA (ed.), *Encyclopedia of Biodiversity*, Vol. 3, Academic Press, pp. 121–139.

Stewart FJ, Newton ILG, Cavanaugh CM (2005) Chemosynthetic endosymbioses: Adaptations to oxic–anoxic interfaces. *TRENDS in Microbiology* **13**, 439–448.

Sundaram JK, Rawal V, Clark MY (2015) Ending malnutrition — from commitment to action. New Delhi, India: FAO & Tulika Books, 175p.

Suttle CA (2007) Marine viruses — major players in the global ecosystem. *Nature Reviews Microbiology* **5**, 801–812.

Tacon AGJ (2001) Increasing the contribution of aquaculture for food security and poverty alleviation. In: Subasinghe R. P., Bueno P. B., Phillips M. J., Hough C., McGladdery S. E. and Arthur J. R., (eds.), *Aquaculture in the Third Millennium. Technical Proceedings of the Conference on Aquaculture in the Third Millennium*, Bangkok, Thailand, 20–25 February 2000, pp. 63–72. NACA, Bangkok, Thailand and FAO, Rome, Italy.

Tacon AGJ, Metian M (2018) Food matters: Fish, income, and food supply — a comparative analysis. *Reviews in Fisheries Science and Aquaculture* **26**, 15–28.

Takami H, Inoue A, Fuji F, Horikoshi K (1997) Microbial flora in the deepest sea mud of the Mariana Trench. *FEMS Microbiology Letters* **152**, 279–285.

Takami H, Kobata K, Nagahama T, Kobayashi H, Inoue A, Horikoshi K (1999) Biodiversity in deep-sea sites located near the south part of Japan. *Extremophiles* **3**, 97–102.

Taylor M (2015) High Hopes for High Seas! The Blog, Huffington Post Online, Available at: http://www.huffingtonpost.com/dame-meg-taylor/high-hopes-for-high-sea_b_7556618.html.

Teske A, Brinkhoff T, Muyzer G, Moser DP, Rethmeier J, Jannasch HW (2000) Diversity of thiosulfate-oxidizing bacteria from marine sediments and hydrothermal vents. *Applied and Environmental Microbiology* **66**, 3125–3133.

Trombetta MJ (2011) Rethinking the securitization of the environment: Old beliefs, new insights. In: Balzacq T (ed.), *Securitization Theory: How Security Problems Emerge and Dissolve*. London: Routledge, pp. 135–149.

UK Government Office for Science (2018) https://www.gov.uk/government/organisations/government-office-for-science.

United Nations Conference on Sustainable Development (2012) Blue Economy Concept Paper, s.l., Rio de Janeiro, Brasil.

UN DESA (United Nations Department of Economic and Social Affairs) (2017) World population prospects. The 2017 revision. Key findings and advance tables. United Nations, New York, USA, 46.

UNESCO (2004) Submarine groundwater discharge: Management implications, measurements and effects. IHP-VI, Series on Groundwater No. 5, IOC Manuals and Guides No. 44.

Webster PJ, Holland GJ, Curry JA, Chang H-R (2005) Changes in tropical cyclone number, duration, and intensity in a warming environment. *Science* **309**, 1844–1846.

Zhang Y, Dong S, Wang H, Tao S, Kiyama R (2016) Biological impact of environmental polycyclic aromatic hydrocarbons (ePAHs) as endocrine disruptors. *Environmental Pollution* **213**, 809–824.

Zhang C, Dang H, Azam F, Benner R, Legendre L, Passow U, Polimene L, Robinson C, Suttle CA, Jiao N (2018) Evolving paradigms in biological carbon cycling in the ocean. *National Science Review* **5**, 481–499.

Ziccardi LM, Edgington A, Hentz K, Kulacki KJ, Driscoll SK (2016) Microplastics as vectors for bioaccumulation of hydrophobic organic chemicals in the marine environment: A state-of-the-science review. *Environmental Toxicology and Chemistry* **35**, 1667–1676.

Chapter 2

Ocean Acidification
Human Activities Increase Atmospheric CO_2, Which Diffuses in the Surface Ocean Decreasing Seawater pH

"During Germany's Presidency of the G7 in 2015, we reiterated our joint willingness to move towards decarbonisation as industrialised countries. I firmly believe that the industrialised countries need to play a very special and major role because they have the capacity to develop the necessary technological innovations that could set standards, but of course also because they have a historical responsibility and have been a significant factor in the rise of CO_2 emissions worldwide."

— Speech by Federal Chancellor Dr. Angela Merkel at COP23 in Bonn on 15 November 2017.

Introduction

The link between the increase in carbon dioxide (CO_2) and the warming of Earth has been at the center of scientific and political debate. At the 2015 United Nations Climate Change Conference held in Paris, of the 197 Parties to the Convention, 180 ratified to agree on a strategy to reduce emissions of CO_2 and other greenhouse gases,[1] with the aim to limit the continuing rise in global temperature to under 2 °C (Anderson and Peters, 2016).

---

[1] Greenhouse gases, such as CO_2 and methane, are gases in the atmosphere that absorb and emit radiant energy within the thermal infrared range, making the Earth warmer.

In addition to warming as a direct result of increases in atmospheric CO_2 levels caused by human activities, the ocean is undergoing shifts in inorganic carbon chemistry (The Royal Society, 2005). Also, these chemical alterations driven by an increase in CO_2 levels in the ocean cause a decrease in the pH of seawater, making the oceans "more acidic" (or rather less alkaline), hence the term "ocean acidification (OA)"[2] (Caldeira and Wickett, 2003). The accelerated acidification of the ocean has been caused by many factors that have contributed to the increased atmospheric CO_2 levels since the beginning of the Industrial Revolution. Among these factors are the excessive consumerism by an exponentially increasing human population, the resulting industrial production and waste, the intimate coupling between the global economy and fossil fuels, and the lack of a sense of urgency in modifying human behavioral patterns and industrial practices to cut down CO_2 emissions. In addition to the increasing demand for energy and goods, the reluctance or inertia of some governments to invest in energy-clean technologies is causing excessive accumulation of carbon in Earth's reservoirs, including the ocean. The ongoing accumulation of CO_2 in the upper ocean and the consequent changes of inorganic carbon chemistry are affecting the health of marine ecosystems (Albright et al., 2010), the abundance and diversity of marine populations (Kroeker et al., 2013), and the capacity of the oceans to take up carbon (Sabine et al., 2004). This chapter brings together the state of our understanding of the chemistry of OA, the impacts of OA on marine animals and plants, the combined effects of OA and other climate-driven stressors, and the threats to the maintenance of ecosystem diversity and function, with a focus on organisms that produce calcium carbonate.

CO_2 Measurements Before Awareness of Ocean Acidification

Although she is rarely given recognition, Eunice Foote, was the first to demonstrate the absorption of heat from solar radiation by CO_2 and water vapour, and she also hypothesized that changes in CO_2 and water vapour in the atmosphere likely govern climate (Foote, 1856). Like in most societies

---

[2]Ocean pH remains and is predicted to be alkaline according to worst-case scenarios. The term "ocean acidification" makes reference to the process of pH decrease toward acidic values rather than the ocean pH itself becoming acidic (even worse-case scenarios forecast ocean pH values above 7).

that have excluded women from certain scientific forums, Foote's results were presented at the Annual Meeting of the American Association for the Advancement of Science not by her, but by Joseph Henry (the then founding director of the Smithsonian Institution) (Jackson, 2019). There were later explorations of Foote's discovery although her work remained largely uncited (e.g., see Tyndal, 1959).

In 1958, Charles D. Keeling started measuring atmospheric CO_2 concentrations from the Mauna Loa Observatory on the Big Island of Hawaii to later reveal the link between fossil fuel combustion and global climate change (Keeling et al., 2005). The Keeling Curve (Figure 1) has become the iconic figure that illustrates CO_2 increases as a result of human releases of fossil fuels through a monitoring program that continues today.

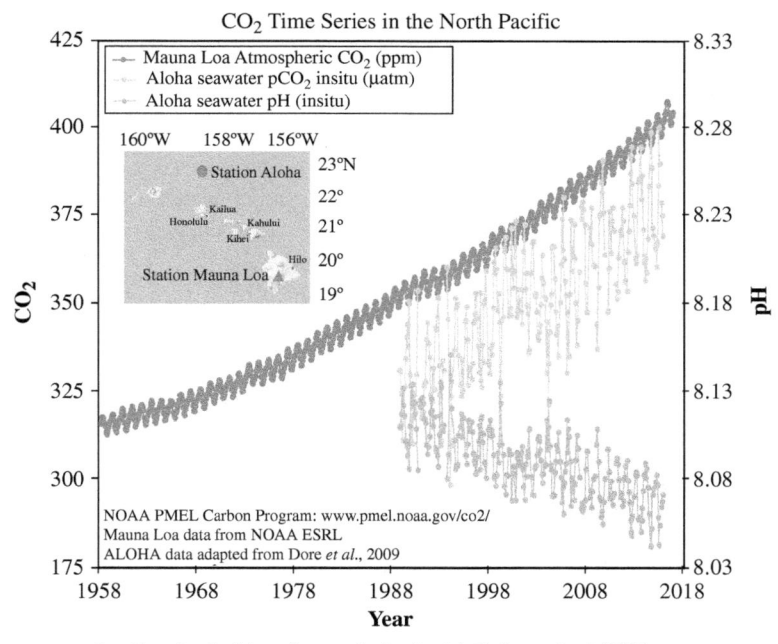

Figure 1. The rise in atmospheric CO_2 at Mauna Loa and seawater chemical data from nearby ocean monitoring station ALOHA, showing a rise in seawater pCO_2 and decrease in seawater pH.
Source: Updated from Feely (2008).

A big challenge has been to make regular, precise, and accurate measurements of CO_2. Keeling started measuring CO_2 in air and water samples every few hours throughout the day and night with a gas manometer, controlled for temperature, pressure, and volume, down to a precision of 0.1% (The American Chemical Society, https://www.acs.org/content/acs/en/education/whatischemistry/landmarks/keeling-curve.html). He noticed that CO_2 concentrations increased at night and decreased during the day. Keeling presented his findings to the Weather Bureau's Division of Meteorological Research that was planning to start a monitoring program of atmospheric CO_2 at remote locations to establish a baseline of CO_2 concentrations. Keeling proposed to deploy a new analytical tool — an infrared gas analyzer — to perform continuous measurements of CO_2 in air samples.

In March 1958, an analyzer was installed at the Weather Bureau's Mauna Loa Observatory, and the first reading was 313 parts per million (ppm) CO_2. After over one year of measurements, Keeling noticed the effect of photosynthesis reducing atmospheric CO_2 levels during the months when photosynthesis was more pronounced (longer days, more intense sunlight). Keeling was also able to estimate that ~50% of all CO_2 released by coal, oil, and natural gas remained in the atmosphere, thus causing CO_2 to increase over time. The remaining ~50% is dissolved into seawater, taken up by plants, or accumulated in soils.

In 1988, a hydrographic time series station, The Hawaii Ocean Time (HOT), series monitoring program was established as part of the Joint Global Ocean Flux Study Program series. Monthly cruises to Station ALOHA, located ~100 km north of Oahu (Figure 1) provided hydrographic, biological, and chemical measurements throughout the water column, revealing that increases in atmospheric CO_2 were mirrored by rises in seawater CO_2 (Figure 1). The availability of this monitoring time series and parallel programs led to the development of sensors and improved protocols and instrumentation to advance our capability to measure carbonate chemistry including CO_2 in seawater. Today, a number of volunteer observing ships, buoys and autonomous systems (coastal, open ocean, coral reef), and hydrographic cruises are mapping out the extent of OA globally. The OA community is a great example of a global working effort to understand a global problem, i.e., how the increases in atmospheric CO_2 are altering the oceans carbon chemistry and the extent to which the oceans will continue to absorb anthropogenic CO_2.

The Origin of Human-Induced OA: Trends in Atmospheric CO_2 During the Anthropocene

The term "Anthropocene" was coined by Paul Crutzen (2006), and it refers to the geological epoch starting from around the time when James Watt designed the steam engine in England. In addition to the technological progress and industrial innovation of the Industrial Revolution, which changed society radically, the sanitary revolution in the late 19th century and medical and public health progress in the 20th century prolonged life expectancy in humans and led to a decline in infant mortality. The resulting epidemiological transition[3] caused a surge in human population (10-fold increase over the last three centuries) (Riddley, 2010; Rappuoli et al., 2011).

Largely as a result of mass industrial processes and transportation to meet human demands for expansion in a globalized economy, the partial pressures[4] of greenhouse gases, including CO_2, have been increasing in the atmosphere since the start of industrialization. The partial pressure of CO_2, a product of many industrial processes such as cement production, burning of fossil fuels like coal, oil, and natural gas, and global deforestation has increased from ~280 parts per million[5] volume[6] (ppmv or ppm) 250 years ago to ~414 ppm in June 2019 (https://scripps.ucsd.edu/programs/keelingcurve/). Within a time frame of months, these increases in atmospheric CO_2 result in equivalent CO_2 levels in the upper mixed layer of the ocean, which have direct consequences on the physiological performance, survival, and functional properties of marine animals and plants. Although the current CO_2 partial pressure in seawater is not unprecedented in the history of the ocean, the rate of change is perhaps faster than observed at any time within the past 65 million years and possibly 300 million years (Hönisch et al., 2012). What follows is a description of the chemistry of OA and its effects on marine ecosystems.

---

[3] The complex change in trends of health and disease and the interactions between these patterns and their demographic, economic, and sociologic determinants and consequences.
[4] The hypothetical pressure of a gas if it alone occupied the volume of a gas mixture at the same temperature.
[5] ppm refers to parts per million, a unit representing the part of a whole number in units of 1/1,000,000.
[6] ppmv refers to parts per million volume, a subunit of ppm that is used for volumes (e.g., mL/m^3).

Basic Chemistry of OA

The carbonate system of seawater is a weak acid–base system composed of dissolved inorganic carbon (DIC) largely as free aqueous carbon dioxide (CO_2), bicarbonate ions (HCO_3^-), and carbonate ions (CO_3^{2-}). In present-day open-ocean waters, a very small percentage of the DIC is present as carbonic acid (H_2CO_3) (Zeebe and Wolf-Gladrow, 2001), which originates from the dissolution of CO_2 into seawater. Because the ocean surface is in direct contact with the atmosphere, any changes in the atmospheric composition of gases will impact directly the concentrations of these gases in the surface ocean (Figure 2).

Under thermodynamic equilibrium, the relationship between atmospheric gaseous CO_2 [$CO_2(g)$] and the concentration of CO_2 in seawater

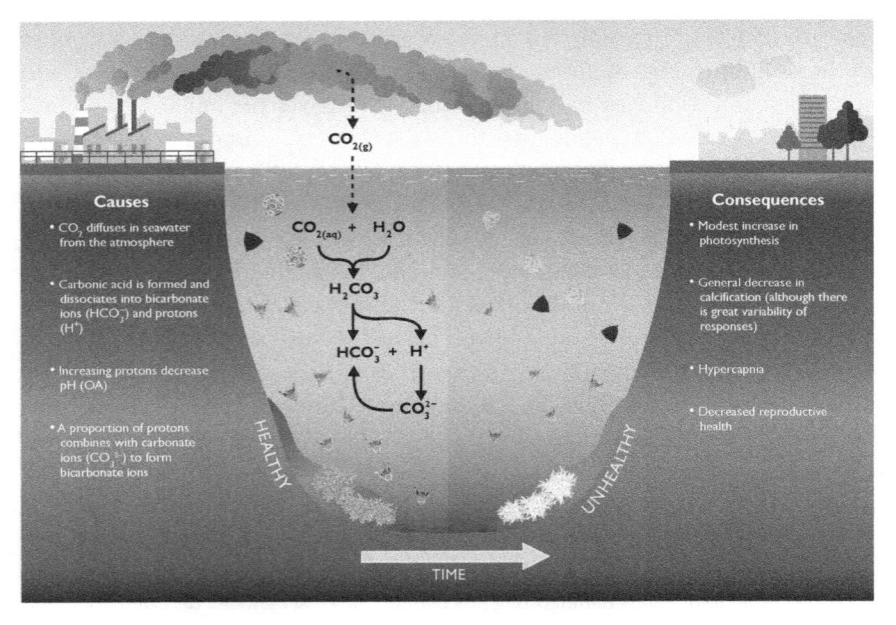

Figure 2. The basic chemistry of OA caused by increasing partial pressure of atmospheric CO_2 [$CO_2(g)$] and consequences on marine ecosystems. The image depicts different types of planktonic organisms. Diatoms (triangular shapes) appear to be favored under OA scenarios (Valenzuela et al., 2018) whereas coccolithophores (represented by round shapes) appear to calcify less in many ocean regions (see Krumhardt et al., 2019). Development of embryos and larvae (blue and yellow symbols representing sea urchin larvae) is negatively affected by OA (see Przeslawski et al., 2015).

[CO$_2$] is defined by Henry's law:

$$CO_2(g) \xrightleftharpoons{K_0} [CO_2] \qquad (1)$$

where K_0 is the solubility coefficient of CO$_2$ in seawater, which is dependent on temperature and salinity (Weiss, 1974), CO$_2$(g) is gaseous CO$_2$, and [CO$_2$] is free aqueous CO$_2$. The dissolved carbonate species are related by the following equilibria:

$$CO_2 + H_2O \xrightleftharpoons{K_1^*} HCO_3^- + H^+ \xrightleftharpoons{K_2^*} CO_3^{2-} + 2H^+ \qquad (2)$$

where K_1^* and K_2^* are the stoichiometric dissociation constants of carbonic acid in seawater (Prieto and Millero, 2001). At typical open-ocean surface seawater pH of 8.2, the speciation between CO$_2$, HCO$_3^-$, and CO$_3^{2-}$ is 0.5%, 89%, and 10.5%, respectively, most of the DIC being in the form of HCO$_3^-$ (Zeebe and Wolf-Gladrow, 2001) (Figure 3). However, these equilibria are being altered as anthropogenic CO$_2$ continues to accumulate in the atmosphere. Specifically, atmospheric CO$_2$ gas diffuses into seawater, a portion of it dissolves in seawater to form carbonic acid (CO$_2$ + H$_2$O ↔

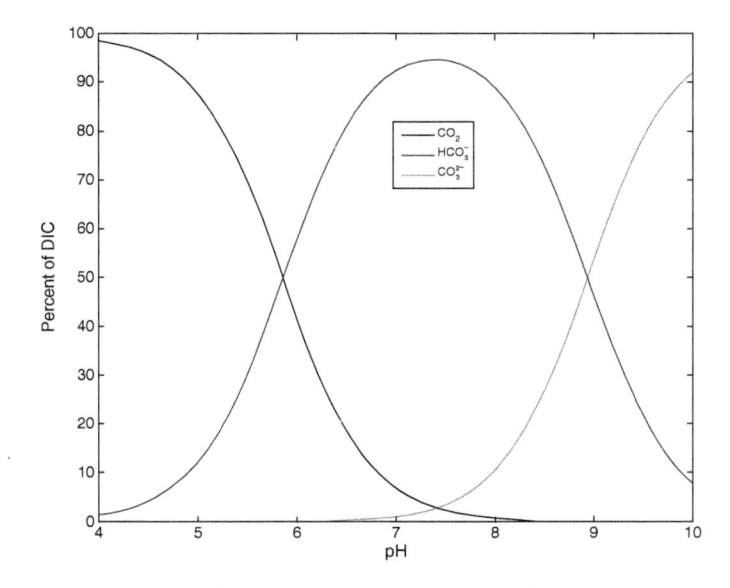

Figure 3. Bjerrum plot illustrating the concentrations of the DIC molecules of the carbonate system in seawater. Scenario for DIC (CO$_2$+HCO$_3^-$+CO$_3^{2-}$) concentration = 2,000 μmol kg^{-1}, temperature $T = 15$ °C, salinity $S = 35$, and pressure $P = 1$ atm. *Source*: From Zeebe and Wolf-Gladrow (2001).

Table 1. An example of carbonate chemistry parameters measured in laboratory experimentation with phytoplankton under conditions representative of past and future climate scenarios.

Parameter	Preindustrial	2035	2100
pCO_2 (ppmv)	280	490	750
$[CO_2]$ (μmol kg^{-1})	10	16	25
$[CO_3^{2-}]$ (μmol kg^{-1})	220	160	110
[DIC] (μmol kg^{-1})	1910	2020	2030
$[HCO_3^-]$ (μmol kg^{-1})	1675	1840	1900
Ω_{cal}	5.3	3.8	2.6
pH	8.15	7.96	7.79
Alkalinity (μeq Liter^{-1})	2220	2227	2160

Note: Ω_{cal} = saturation state of calcite (a polymorph of $CaCO_3$).
Source: From Iglesias-Rodriguez et al. (2008).

H_2CO_3), and the H_2CO_3 dissociates into HCO_3^- and protons ($H_2CO_3 \leftrightarrow HCO_3^- + H^+$) (Figure 2). As a result of the increase in the concentration of H^+, the pH of seawater decreases, hence the term "ocean acidification" (Caldeira and Wickett, 2003). This increase in protons shifts the equilibrium between HCO_3^- and CO_3^{2-} toward HCO_3^- (see Equation (2)). Thus, OA, driven by increasing anthropogenic CO_2 levels in seawater, is characterized by increases in CO_2 and HCO_3^- and decreases in pH and CO_3^{2-} ion concentrations relative to preindustrial conditions (Table 1, Figure 2).

An interesting parameter often measured in OA research is total alkalinity (TA), which is defined as the charge difference between the major cations and anions in seawater. TA remains largely unaltered when CO_2 is added to seawater because the charge balance of seawater stays the same, i.e., the number of positive ions equals the number of negative ions generated by CO_2 entering seawater from the atmosphere.

Impacts of OA on Marine Ecosystem Processes

The pH and relative abundances of inorganic carbon species in seawater are altered by human-driven OA and other abiotic (physical and chemical) and biotic processes that produce or consume H^+. Any change in pH, including that caused by OA, will result in shifts in the $HCO_3^-:CO_3^{2-}:CO_2$ ratio. This ratio is important for marine animals and plants because the different inorganic carbon species are substrates or products of important physiological processes that affect their health, behavior, and survival, and the $HCO_3^-:CO_3^{2-}:CO_2$ ratio also determines the biogeochemical properties of ecosystems. For example, changes in levels of CO_2, the substrate for photosynthetic carbon fixation, can alter photosynthetic rates; changes in HCO_3^- can impact photosynthesis and calcification because HCO_3^- is an indirect source of carbon for both processes (inside cells, HCO_3^- is converted into CO_2 and CO_3^{2-}, the substrates for photosynthesis and calcification, respectively); and the concentration of CO_3^{2-} determines the fate of $CaCO_3$ because low concentrations of CO_3^{2-} (corrosive waters) can result in $CaCO_3$ dissolution. Additionally, other climate-driven stressors can alter (sometimes magnify) all these biological effects; thus, OA needs to be studied in the context of other anthropogenically driven stressors, such as warming, shifts in nutrients/resources, and hypoxia/anoxia.

Effect of OA on photosynthesis

A rise in CO_2 may be an evolutionary advantage for marine photosynthetic organisms because photosynthesis is limited by the diffusion of CO_2 in seawater, the slow dehydration kinetics of bicarbonate to CO_2 ($HCO_3^- + H^+ \rightarrow CO_2 + H_2O$), and the low affinity of ribulose-1,5-bisphosphate carboxylase/oxygenase (RubisCO) for its inorganic carbon substrate, CO_2 (Reinfelder, 2011). This is why the vast majority of eukaryotic[7] phytoplankton and macroalgae have developed a carbon concentrating mechanism (CCM) (Reinfelder, 2011), i.e., cells have acquired, over evolutionary timescales, strategies to take up inorganic carbon species other than CO_2, e.g., HCO_3^-. Also, most aquatic photosynthetic organisms have membrane-bound enzymes and transport mechanisms that facilitate inorganic carbon uptake, for example, carbonic anhydrase on the outer layers of the plasma membrane as well as inside cells, and HCO_3^- transporters, enhancing inorganic carbon acquisition (Reinfelder, 2011).

When studying OA, one often forgets that the pH of seawater is "less acidic" (more alkaline) than the pH inside cells. Typically, the mean pH in the cytosol, nucleoplasm, mitochondria, and plastid stroma of cells is between 7.1 and 7.8 (Raven and Smith, 1980; Burns and Beardall, 1987; Anning et al., 1996; Venn et al., 2009), which is significantly lower

---

[7]Eukaryotes are organisms (both unicellular and multicellular) with cells that have a defined nucleus separated from the rest of the cell via membranes. In contrast, prokaryotes such as bacteria and cyanobacteria are unicellular organisms that do not have a distinct nucleus with a membrane.

than the average surrounding seawater pH (~8.1). This means that the concentration of CO_2 is actually greater inside cells than in the surrounding seawater. Although an increase in the CO_2 concentration of seawater could potentially be energetically favorable to photosynthetic cells, and particularly those with poor CCMs (Reinfelder, 2011), a decrease in pH also affects membrane potential, energy partitioning, and enzyme activity (Beardall and Raven 2004, Giordano et al., 2005), all having the potential to reduce phytoplankton fitness (Berge et al., 2010). Observations through experimental work suggest that photosynthetic responses to OA are relatively small for most species and highly variable throughout taxa (Koch et al., 2013; Mackey et al., 2015), the suggestion being that the potential photosynthetic benefits of elevated CO_2 are likely to be minor relative to the impact on the cell's energy balance.

Controls on biogenic calcium carbonate precipitation

Perhaps, the most discussed aspect of OA is its impact on marine animals and plants that precipitate calcium carbonate ($CaCO_3$), for example, corals, foraminifera, coccolithophores, pteropods, echinoderms, and other groups that calcify during some stages of their life cycle. In $CaCO_3$-producing organisms (or calcifiers), the carbonate ion (CO_3^{2-}) is the substrate of carbon used for calcification ($Ca^{2+} + CO_3^{2-} \rightarrow CaCO_3$). However, there seems to be a general consensus that the source of DIC from seawater for the calcification process is actually not CO_3^{2-} but bicarbonate ions (HCO_3^-) in most calcifiers, including corals (Marubini and Thake 1999; Furla et al., 2000; Jury et al., 2010) and coccolithophores (Sikes and Wilbur, 1982). The main explanations for the need for a CCM that imports HCO_3^- from the external seawater into the organism and subsequently into the calcification reservoir (Figure 4) are as follows: (a) CO_3^{2-} represents only 15% of the total DIC whereas HCO_3^- is the most abundant DIC species in seawater, and therefore HCO_3^- uptake is ecologically and evolutionarily advantageous, and (b) cells appear to lack a mechanism to import CO_3^{2-} through membranes, and therefore, an alternative source of DIC (i.e., HCO_3^-) must be imported from the seawater into the organism and eventually into the calcification fluid, where it is converted into CO_3^{2-}.

In the calcification fluid (e.g., coccolith vesicle in coccolithophores, calcicoblast in corals), conversion of HCO_3^- into CO_3^{2-} is accompanied by the release of protons ($HCO_3^- \rightarrow CO_3^{2-} + H^+$). Given that the accumulation of H^+ in the calcification reservoir causes acidification, which would lead

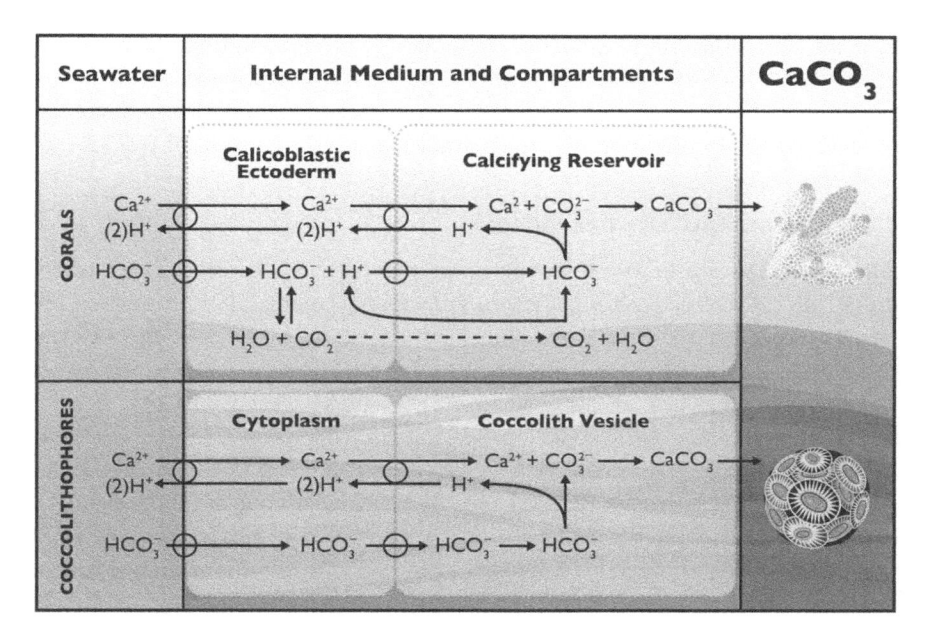

Figure 4. Examples of inorganic carbon acquisition in two types of organisms that precipitate CaCO3: corals and coccolithophores. The circles represent active transport of molecules through membranes. In the case of HCO_3^-, its transport may be facilitated by carbonic anhydrase (which catalyzes the conversion between CO_2, HCO_3^- and H_2O) or by symport (simultaneous import of Na^+ and HCO_3^- through the membrane) or antiport (extrusion of Cl^- concomitant with HCO_3^- uptake through the membrane) mechanisms.

to CaCO3 dissolution, energy is required to extrude protons from the calcification reservoir in order to maintain elevated pH and promote calcification (Taylor et al., 2011). The supply of Ca^{2+}, with concentrations in seawater of approximately two orders of magnitude higher than CO_3^{2-}, can be facilitated by Ca^{2+}-stimulated ATPases (Araki and Gonzalez, 1998; Al-Horani et al., 2003). In corals, a Ca^{2+}/H^+ exchanger mediates the removal of two H^+ and the transport of one Ca^{2+} ion into the calcification reservoir. This process appears to serve both as a supply of Ca^{2+} ions and a homeostatic mechanism in the calicoblast (Cohen and Holcomb, 2009).

The paragraphs above describe how calcifiers acquire carbon and calcium, and maintain calcification by extruding H^+ from the CaCO3 reservoir. Given that OA leads to an increase of HCO_3^- concentrations in seawater, one could assume that calcification can be maintained and even enhanced as the oceans are getting richer in HCO_3^-. This has been observed in a number of studies exploring the effect of OA on different taxa (Ries et al., 2009). However, the stability of $CaCO_3$ is compromised in the ocean as seawater becomes more acidified because concomitant with the increase of HCO_3^- is a decline in CO_3^{2-} that could lead to dissolution of $CaCO_3$.

Controls on $CaCO_3$ dissolution

The concentration of Ca^{2+} in the open ocean tends to remain unaltered unlike in coastal ecosystems, where terrigenous material that washes off into the sea can increase Ca^{2+} concentrations. However, the decrease in CO_3^{2-} caused by OA can compromise the stability of $CaCO_3$. Because the concentration of Ca^{2+} is >50 times the concentration of CO_3^{2-}, Ca^{2+} is not rate-limiting for $CaCO_3$ precipitation. Therefore, the progressive decline of CO_3^{2-} concentrations in seawater due to OA can result in shell dissolution ($CaCO_3 \rightarrow Ca^{2+} + CO_3^{2-}$) in calcifying organisms under conditions considered "corrosive", i.e., when the concentrations of CO_3^{2-} are very low. The degree of "corrosiveness" is typically quantified in terms of the saturation state (Ω) of $CaCO_3$:

$$\Omega = [Ca^{2+}][CO_3^{2-}]/K_{sp}^* \tag{3}$$

where K_{sp}^* is the apparent solubility product for $CaCO_3$. There are many polymorphs of $CaCO_3$, for example, calcite (produced by coccolithophores and foraminifera), high-Mg calcite (e.g., in echinoderms), aragonite (produced by corals and pteropods), amorphous $CaCO_3$ (e.g., in early stages of sea urchin development), and the less common vaterite. Indeed, in the marine environment, a wide range of planktonic (e.g., coccolithophores, foraminifera, pteropods, larval fish) and benthic (e.g., green and red algae, bivalves, gastropods, corals, echinoderms, crustacea, foraminifera, serpulid worms, bryozoa, sponges) groups produce $CaCO_3$ predominantly as calcite, aragonite, or "high-Mg calcite" ($\%MgCO_3 > 4$), each with distinct susceptibilities to dissolution (Morse et al., 2006; Lebrato et al., 2010). These polymorphs of $CaCO_3$ share the same stoichiometry, but they differ in their atomic configurations (Morse et al., 2006). Due to these structural differences, calcite is more stable in seawater than aragonite, and high-Mg calcite is more susceptible to dissolution than aragonite.

Controversy

Perhaps, the most controversial issue at the center of the debate about OA is its effect on calcification. Most studies that have assessed the impacts of OA on net calcification have found a decrease in calcification. However, as taxonomic coverage has increased, more variation has been encountered in responses to OA. Studies have revealed trends varying from decreased calcification (Riebesell et al., 2000; Zondervan et al., 2001; Orr et al., 2005; Hoegh-Guldberg et al., 2007; Anthony et al., 2008), increased calcification (Iglesias-Rodriguez et al., 2008; Shi et al., 2009; Uthicke et al., 2016), sometimes at the expense of some metabolic cost, such as a decrease in biomass (Wood et al., 2008; Kroeker et al., 2010); and neutral and non-uniform responses (Langer et al., 2006, 2009; Rodolfo-Metalpa et al., 2011).

How do we explain this diversity of responses of a process like calcification that is widespread and well understood thermodynamically? On the one hand, the increase in seawater HCO_3^- levels (bicarbonation) can maintain or even facilitate calcification because calcifiers import bicarbonate (the concentration of which increases with OA) into their calcification reservoirs. On the other hand, a decrease in the concentration of carbonate ions can cause $CaCO_3$ dissolution. One explanation for the apparent neutral changes and sometimes increase in net calcification of some calcifiers exposed to OA may be bioprotection of the carbonate biomineral. Many calcifiers can build organic layers that cover $CaCO_3$ and thus prevent $CaCO_3$ dissolution. This organic layer has been observed in coccolithophores, crustacea (epicuticle), echinoids, mollusks (periostracum), corals (ectoderm), and calcifying algae (utricles) (Young and Henriksen, 2003; Allemand et al., 2004; Marin et al., 2008; Ries et al., 2009). The observed decline in calcification may be particularly important in species whose carbonate shells/plates are directly exposed to seawater and not protected by organic layers (Ries et al., 2009; Tunnicliffe et al., 2009).

Protective organic coverings, along with efficient proton pumping, appear to be the most important factors controlling the calcification response of marine organisms to CO_2-induced OA. The extent to which organisms can buffer acidification and maintain calcification rates inside their calcification reservoir will be largely dependent on the efficiency of transmembrane ion pumps that maintain pH and the concentrations of DIC species (CO_2, CO_3^{2-}, HCO_3^-) and Ca^{2+} at levels conducive to calcification. Specifically, efficient proton pumps are required to extrude H^+ in order to prevent acidification of the calcification fluid (Taylor et al., 2011; Al-Horani et al., 2003), which could lead to $CaCO_3$ dissolution ($CaCO_3 + H^+ \rightarrow Ca^{2+} + HCO_3^-$). The source of the variable patterns observed in some species of echinoids, crustaceans, calcareous green and coralline red algae, coccolithophores, molluscs, and corals may be inherent differences in the efficiencies of their proton pumps (Ries, 2011) and shifts in the alkalinity-pumping regimes driven by alterations in seawater conditions (Gagnon, 2013).

In addition to the conditions that enhance (increase in [HCO_3^-]) and suppress (decrease in [CO_3^{2-}]) $CaCO_3$ precipitation, there are other effects of OA on organism physiology. Specifically, a decrease in pH can be detrimental to organisms and could potentially lead to the suppression of metabolic processes other than calcification. For example, an increase of CO_2 in the blood of marine animals, also known as hypercapnia, can be extremely detrimental and potentially cause death. Some organisms can take up bicarbonate to compensate for the decrease in pH, and organisms that can concentrate bicarbonate by factors of 4–10 relative to control conditions are generally more tolerant of hypercapnia, whereas those with limited potential to accumulate bicarbonate might be more vulnerable (Heisler, 1993). Therefore, in addition to $CaCO_3$ dissolution effects, there are physiological repercussions of OA that could cause a decrease in ecological fitness and be detrimental to the maintenance of population diversity and function.

Effect of OA on the Reproductive Health of Marine Organisms

A number of studies have shown a detrimental effect of OA on animal reproduction (e.g., Kurihara, 2008; Dupont et al., 2010), altered transition between life-cycle stages (Dupont et al., 2013), and transgenerational plasticity whereby acclimation to environmental pressure takes place via changes in morphology, physiology, or life histories across generations (Parker et al., 2012).

Measurements of performance have indicated that early stages of the life cycle of an organism are generally more sensitive to OA than adult stages (Kurihara and Shirayama, 2004; Ross et al., 2011). Larvae are particularly susceptible to OA during stages that involve the formation of an early skeleton (for example, spicules in echinoderm larvae), which forms during gastrulation, between 8 and 24 h after fertilization (Killian and Wilt, 2008). In many echinoderms, the result of exposure to low-pH waters is thin or corroded spicules, which can be detrimental to population fitness because spicules have a role in larval movement, feeding, and settling. The impairment of any of these activities, such as settling onto a substrate, can delay the transition of early larvae into more mature larval stages and make larvae more susceptible to predation (Ross et al., 2011). In some but not all animals, OA has also been found to affect sperm motility, which has been used as an indicator of male fitness. For example, OA affects sperm motility and swimming speed in sea urchins (Havenhand et al., 2008), although no effects have been found in oysters (Havenhand and Schiegel, 2009). In addition to these physiological effects, many parts of the echinoderm (e.g., skeletal rods, adult tests, teeth, and spicules of larvae and adults) are composed of amorphous calcium carbonate, a $CaCO_3$ polymorph that is extremely susceptible to dissolution (Politi et al., 2004). Therefore, organisms bearing this and other polymorphs of $CaCO_3$ that are less stable in seawater than calcite (e.g., high-Mg calcite, aragonite, and vaterite) are particularly vulnerable to OA if their calcified structures undergo dissolution.

Naturally "Acidified" Marine Ecosystems: Volcanic CO_2 Vents and Upwelling Zones

Submarine CO_2 vents are marine ecosystems undergoing chronic exposure to elevated CO_2, and thus they provide a platform to test how organisms survive, perform, and interact in naturally "acidified" conditions (Hall-Spencer et al., 2008; Bianchi et al., 2011; Fabricius et al., 2011; Inoue et al., 2013; Pettit et al., 2013). Although most submarine CO_2 vents that have been studied are shallow (<5 m depth), some studies have explored hydrothermal vents at greater depths, down to a few hundred meters (e.g., Bianchi et al., 2011; Pettit et al., 2013). Studies of diversity, productivity, and trophic structure in marine communities have revealed major declines in calcifying organisms and increased abundance of macroalgae, seagrasses, or soft corals (Hall-Spencer et al., 2008; Kroeker et al., 2011, 2012). These results indicate that these low-pH ecosystems select for non-calcareous marine life (e.g., Hall-Spencer et al., 2008; Fabricius et al., 2011; Kroeker et al., 2011; Inoue et al., 2013; Kroeker et al., 2013; Pettit et al., 2013), although some calcifying organisms such as calcareous invertebrates can be abundant at intermediate pH values (pH 7.41–7.99) (Cigliano et al., 2010). However, most vents are imperfect analogues of OA because, in addition to low pH, many seafloor vent ecosystems have elevated temperatures and many have high concentrations of toxic metals (e.g., cadmium, silver, strontium, barium) (Van Dover, 2000). Therefore, the organisms that succeed in many of these environments have developed adaptations to not only OA but also other stressors.

Another natural phenomenon that causes seasonal OA is upwelling. If we visualize a column of water in the ocean, the concentration of CO_2

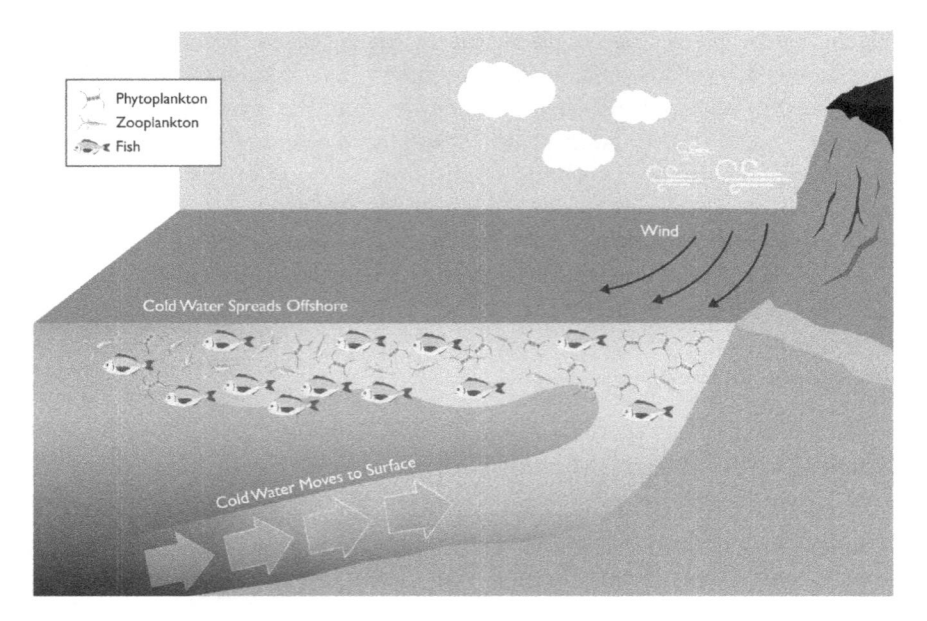

Figure 5. Schematic representation of upwelling. Wind-driven cool (large blue arrows) deep water rich in nutrients and carbon rises to the surface during upwelling. The fertilization of the surface coastal waters promotes primary production (phytoplankton growth), providing food for consumers (zooplankton). As a result, upwelling zones are responsible for the most important fisheries in the world.

typically increases with depth. This gradient is primarily driven by bacterial respiration and the solubility pump. The latter reflects the fact that the solubility of CO_2 is an inverse function of temperature and pressure, which control the extent to which gases (including CO_2) dissolve in seawater. During upwelling, wind blowing across the surface ocean causes dense, cold, nutrient-rich and CO_2-rich seawater to rise from depth to the ocean surface (Feely et al., 2008) (Figure 5). Rather than a chronic state, this is a seasonal and episodic phenomenon, and organisms inhabiting upwelling zones are adapted to seasonal rather than permanent exposure to acidification and cooling. Because upwelled waters are rich in nutrients, upwelling regions are among the most productive ecosystems in the world's oceans (Pauly and Christensen, 1995). Specifically, concentrations of important nutrients such as phosphate and nitrate can increase severalfold during upwelling (van Geen et al., 2000), and this increase typically enhances productivity (Brock et al., 1991; Brock and McClain, 1992; Brzezinski and Washburn, 2011). In some upwelling regions, such as the northeastern Pacific Ocean,

$p\text{CO}_2$ values can exceed 1,000 ppm near surface waters (Feely et al., 2008). As a result, these waters can be corrosive to carbonate minerals (Feely et al., 2008; Hauri et al., 2013). Studying these ecosystems is important and timely because over the last 150 years, upwelling has intensified (Gutiérrez et al., 2011). Also, it has been proposed that anthropogenic greenhouse forcing will continue to enhance coastal upwelling by intensifying the alongshore wind stress on the ocean surface in eastern boundary current regions (Bakun, 1990; Gutierrez et al., 2011). This means that the west coast of continents where upwelling occurs is likely undergoing more intense or prolonged acidification.

Combined Effects of OA and Other Climate Stressors on Marine Organisms

A major challenge in predicting the effect of climate change on marine ecosystems is that stressors caused directly or indirectly by human activities such as OA do not occur in isolation, and other factors such as warming (Gleckler et al., 2012; Acosta Navarro et al., 2016), hypoxia/anoxia (Keeling et al., 2010), pollution (Walker and Livingstone, 1992; Shi et al., 2016), light (Donahue et al., 2019), and alterations in nutrient cycling (Greaver et al., 2016) can add to or counteract the effect of OA (Gunderson et al., 2016).

Among the most vulnerable ecosystems with respect to OA are coral reefs because many tens of thousands of species depend on their structural complexity and the services provided by the corals' carbonate skeletons (Kleypas et al., 1999; Hoegh-Guldberg et al., 2007; Fabricius et al., 2011), which are affected by OA. But coral reefs are also being impacted by other stressors including coastal eutrophication and heat waves, which cause devastation in coral ecosystems (see Chapter 3). Another example of how combined stressors could impact ecosystems can be seen in a field study at a CO_2 seep off Papua New Guinea. This study revealed that the combination of OA and temperature stress is likely to severely reduce the diversity, structural complexity, and resilience of Indo-Pacific coral reefs within this century (Fabricius et al., 2011). However, complex and varying responses to different combinations of multiple stressors means that the interpretation of results is not always straightforward. For example, in lab experiments with the coccolithophore E. huxleyi, the physiological trends observed using CO_2 levels representative of different climate scenarios varied depending on other factors including nutrient availability (Rouco et al., 2013).

In echinoderms, several responses have been found, including the increase in the growth of adult sea stars under warming and acidification (Gooding et al., 2009). On the other hand, warming appears to dramatically reduce gastrulation success and impair early development of a sea urchin while no major effect on early developmental stages was found under elevated CO_2 (Byrne et al., 2009). Feedback effects, such as the sensitivity of obligatory symbionts to heat stress, could contribute to and further complicate organism responses, particularly in corals (Hoegh-Guldberg et al., 2007).

Key Points

- The levels of CO_2 in the atmosphere and the surface ocean have been increasing consistently since the start of the Industrial Revolution from ~280 ppm to levels exceeding 400 ppm in 2019.
- Although the levels of CO_2 are not unprecedented in recent Earth's history, the rate of CO_2 increase since the start of the Industrial Revolution is the fastest Earth has witnessed for at least the past 65 million years.
- As a result of OA, the concentrations of CO_2 and HCO_3^- continue to increase, whereas pH and the concentration of CO_3^{2-} are declining.
- OA will likely have a modest impact on photosynthesis, with small increases or neutral responses in photosynthetic carbon fixation.
- OA will cause a decrease in the concentration of CO_3^{2-}, which in turn can lead to dissolution of carbonate plates, shells, tests, and reefs. However, some calcifiers are able to maintain and even enhance $CaCO_3$ production under simulated OA conditions, although often at the expense of some physiological cost like growth.
- OA appears to have a detrimental effect on reproduction and development of marine animals. Hotspots for OA such as deep sea vents or upwelling zones provide unique opportunities to test biological responses and assess which organisms are selected against and which are favored in these environments.
- Attention must be paid to OA in the context of other environmental stressors. For example, in addition to OA, other climate-driven stressors such as warming can magnify or suppress the impacts of OA and the physiological acclimations thereto.

Questions

1. What molecules of the dissolved inorganic carbon system increase in seawater with ocean acidification?
2. Are the current CO_2 levels in the atmosphere the highest in Earth's history?
3. Indicate the direction of the arrows in the following seawater equilibria under increasing atmospheric CO_2:

$$CO_2 + H_2O \longrightarrow H_2CO_3 \longrightarrow HCO_3^- + H^+ \longrightarrow CO_3^{2-} + H^+$$

4. OA can enhance photosynthetic carbon fixation in some organisms. Which photosynthetic organisms will profit more from an increase in CO_2 concentrations in seawater: those with or without a CCM? Justify your answer.
5. How does OA driven by increasing anthropogenic CO_2 affect alkalinity in seawater?
6. If the concentration of HCO_3^- increases with ocean acidification, and HCO_3^- is the primary source of DIC from seawater for calcification in marine $CaCO_3$-producing organisms, why do many calcifiers undergo a decrease in calcification with ocean acidification?
7. What is the main inorganic carbon molecule that controls $CaCO_3$ dissolution in seawater?
8. Why are larvae among the most susceptible life-cycle stages to OA during their development into adults?
9. What are examples of naturally occurring OA in marine systems?
10. Give examples of physiological impacts on OA on marine animals.

References

Acosta Navarro JC, Varma V, Riipinen I, Seland Ø, Kirkevåg A, Struthers H, Iversen T, Hansson H-C, Ekman AML (2016) Amplification of Arctic warming by past air pollution reductions in Europe. *Nature Geoscience* **9**, 277–281.

Albright R, Mason B, Miller M, Langdon C (2010) Ocean acidification compromises recruitment success of the threatened Caribbean coral *Acropora palmate*. *Proceedings of the National Academy of Sciences* **107**, 20400–20404.

Al-Horani FA, Al-Moghrabi SM, de Beer D (2003) The mechanism of calcification and its relation to photosynthesis and respiration in the scleractinian coral *Galaxea fascicularis*. *Marine Biology* **142**, 419–426.

Allemand D, Ferrier-Pagès C, Furla P, Houlbrèque F, Puverel S, Reynaud S, Tambutté E, Tambutté S, Zoccola D (2004) Biomineralisation in reef-building corals: From molecular mechanisms to environmental control. *Comptes Rendus Palevol* **3**, 453–467.

Anderson K, Peters G (2016) The trouble with negative emissions. *Science* **354**, 182–183.

Anning T, Nimer N, Merret MJ, Brownlee C (1996) Costs and benefits of calcification in coccolithophorids. *Journal of Marine Systems* **9**, 45–56.

Anthony KRN, Kline DI, Diaz-Pulido G, Dove S, Hoegh-Guldberg O (2008) Ocean acidification causes bleaching and productivity loss in coral reef builders. *Proceedings of the National Academy of Sciences* **105**, 17442–17446.

Araki Y, González EL (1998) V- and P-type Ca^{2+}-stimulated ATPases in a calcifying strain of *Pleurochrysis* sp. (Haptophyceae). *Journal of Phycology* **34**, 79–88.

Bakun A (1990) Global climate change and intensification of coastal ocean upwelling. *Science* **247**, 198–201.

Beardall J, Raven JA (2004) The potential effects of global climate change on microalgal photosynthesis, growth and ecology. *Phycologia* **43**, 26–40.

Berge T, Daugbjerg N, Anderson BB, Hansen PJ (2010) Effect of lowered pH on marine phytoplankton growth rates. *Marine Ecology Progress Series* **416**, 79–91.

Bianchi CN, Dando PR, Morri C (2011) Increased diversity of sessile epibenthos at subtidal hydrothermal vents: Seven hypotheses based on observations at Milos Island, Aegean Sea. *Advances in Oceanography and Limnology* **2**, 1–31.

Brock JC, McClain, CR (1992) Interannual variability of the southwest monsoon phytoplankton bloom in the northwest Arabia Sea. *Journal of Geophysical Research* **97**, 733–750.

Brock JC, McLain CR, Luther ME, Hay WW (1991) The phytoplankton bloom in the Northwestern Arabian Sea during the Southwest Monsoon of 1979. *Journal of Geophysical Research* **96**, 623–620.

Brzezinski MA, Washburn L (2011) Phytoplankton primary productivity in the Santa Barbara Channel: Effects of wind-driven upwelling and mesoscale eddies. *Journal of Geophysical Research* **116**, doi:10.1029/2011JC007397.

Burns BD, Beardall J (1987) Utilization of inorganic carbon by marine microalgae. *Journal of Experimental Marine Biology and Ecology* **107**, 75–86.

Byrne M, Ho M, Selvakumaraswamy P, Nguyen HD, Dworjanyn SA, Davis AR (2009) Temperature, but not pH, compromises sea urchin fertilization and early development under near-future climate change scenarios. *Proceedings of the Royal Society B* **276**, 1883–1888.

Caldeira K, Wickett ME (2003) Anthropogenic carbon and ocean pH. *Nature* **425**, 365.

Cigliano M, Gambi MC, Rodolfo-Metalpa R, Patti FP, Hall-Spencer JM (2010) Effects of ocean acidification on invertebrate settlement at volcanic CO_2 vents. *Marine Biology* **157**, 2489–2502.

Cohen AL, Holcomb M (2009) Why corals care about ocean acidification: Uncovering the mechanism. *Oceanography* **22**, 118–127.

Crutzen PJ (2006) The "Anthropocene". In: Ehlers E., Krafft T. (eds.), *Earth System Science in the Anthropocene*. Springer, Berlin, Heidelberg.
Donahue K, Klaas C, Dillingham PW, Hoffmann LJ (2019) Combined effects of ocean acidification and increased light intensity on natural phytoplankton communities from two Southern Ocean water masses. *Journal of Plankton Research*, doi:10.1093/plankt/fby048.
Dupont S, Ortega-Martinez O, Thorndyke M (2010) Impact of near-future ocean acidification on echinoderms. *Ecotoxicology* **19**, 449–462.
Dupont S, Dorey N, Stumpp M, Melzner F, Thorndyke M (2013) Long-term and trans-life-cycle effects of exposure to ocean acidification in the green sea urchin *Strongylocentrotus droebachiensis*. *Marine Biology* **160**, 1835–1843.
Fabricius KE, Langdon C, Uthicke S, Humphrey C, Noonan S, De'ath G, Okazaki R, Muehllehner N, Glas MS, Lough JM (2011) Losers and winners in coral reefs acclimatized to elevated carbon dioxide concentrations. *Nature Climate Change* **1**, 1–5.
Feely RA (2008) Ocean Acidification. In State of the Climate in 2007, Levinson DH and Lawrimore JH (eds.). *Bull. Am. Meteorol. Soc.* **89**(7), S58.
Feely RA, Sabine CL, Hernandez-Ayon JM, Ianson D, Hales B (2008) Evidence for upwelling of corrosive "acidified" water onto the continental shelf. *Science* **320**, 1490–1492.
Foote E (1856) Circumstances affecting the heat of the Sun's rays. *American Journal of Science and Arts* **22**, 382–383.
Furla P, Galgani I, Durand I, Allemand D (2000) Sources and mechanisms of inorganic carbon transport for coral calcification and photosynthesis. *The Journal of Experimental Biology* **203**, 3445–3457.
Gagnon AC (2013) Coral calcification feels the acid. *Proceedings of the National Academy of Sciences* **110**, 1567–1568.
Giordano M, Beardall J, Raven JA (2005) CO_2 concentrating mechanisms in algae: Mechanisms, environmental modulation, and evolution. *Annual Review of Plant Biology* **56**, 99–131.
Gleckler PJ, Santer BD, Domingues CM, Pierce DW, Barnett TP, Church JA, Taylor KE, AchutaRao KM, Boyer TP, Ishii M, Caldwell PM (2012) Human-induced global ocean warming on multidecadal timescales. *Nature Climate Change* **2**, 524–529.
Gooding RA, Harley CDG, Tang E (2009) Elevated water temperature and carbon dioxide concentration increase the growth of a keystone echinoderm. *Proceedings of the National Academy of Sciences USA* **106**, 9316e9321.
Greaver TL, Clark CM, Compton JE, Vallano D, Talhelm AF, Weaver CP, Band LE, Baron JS, Davidson EA, Tague CL, Felker-Quinn E, Lynch JA, Herrick JD, Liu L, Goodale CL, Novak KJ, Haeuber RA (2016) Key ecological responses to nitrogen are altered by climate change. *Nature Climate Change* **6**, 836–843.
Gunderson AR, Armstrong EJ, Stillman JH (2016) Multiple stressors in a changing world: The need for an improved perspective on physiological responses to the dynamic marine environment. *Annual Review of Marine Science* **8**, 357–378.

Gutiérrez D, Bouloubassi I, Sifeddine A, Purca S, Goubanova K, Graco M, Field D, Méjanelle L, Velazco F, Lorre A, Salvatteci R, Quispe D, Vargas G, Dewitte B, Ortlieb L (2011) Coastal cooling and increased productivity in the main upwelling zone off Peru since the mid-twentieth century. *Geophysical Research Letters* **38**, L07603, doi:10.1029/2010GL046324.

Hall-Spencer, JM, Rodolfo-Metalpa R, Martin S, Ransome E, Fine M, Turner SM, Rowley SJ, Tedesco D, Buia MC (2008) Volcanic carbon dioxide vents show ecosystem effects of ocean acidification. *Nature* **454**, 96–99.

Hauri C, Gruber N, Vogt M, Doney SC, Feely RA, Lachkar Z, Leinweber A, McDonnell AMP, Munnich M, Plattner G-K (2013) Spatiotemporal variability and long-term trends of ocean acidification in the California Current System. *Biogeosciences* **10**, 193–216.

Havenhand JN, Schlegel P (2009) Near-future levels of ocean acidification do not affect sperm motility and fertilization kinetics in the oyster *Crassostrea gigas*. *Biogeosciences* **6**, 3009–3015.

Havenhand JN, Buttler FR, Thorndyke MC, Williamson JE (2008) Near-future levels of ocean acidification reduce fertilization success in a sea urchin. *Current Biology* **18**, R651–R652.

Heisler N (1993) Acid-base regulation in response to changes of the environment: Characteristics and capacity. In: Rankin J. C., Jensen F. B. (eds.), *Fish Ecophysiology*. Chapman & Hall Fish and Fisheries Series, Vol. 9. Springer, Dordrecht.

Hoegh-Guldberg O, Mumby PJ, Hooten AJ, Steneck RS, Greenfield P, Gomez E, Harvell CD, Sale PF, Edwards AJ, Caldeira K, Knowlton N, Eakin CM, Iglesias-Prieto R, Muthiga N, Bradbury RH, Dubi A, Hatziolos ME (2007) Coral reefs under rapid climate change and ocean acidification. *Science* **318**, 1737–1742.

Hönisch B, Ridgwell A, Schmidt DN, Thomas E, Gibbs SJ, Sluijs A, Zeebe RE, Kump L, Martindale RC, Greene SE, Kiessling W, Ries J, Zachos J, Royer DL, Barker S, Marchitto Jr. TM, Moyer R, Pelejero C, Ziveri P, Foster GL, Williams B (2012) The geological record of ocean acidification. *Science* **335**, 1058–1063.

Iglesias-Rodriguez MD, Halloran PR, Rickaby REM, Hall IR, Colmenero-Hidalgo E, Gittins JR, Green DRH, Tyrell T, Gibbs SJ, von Dassow P, Rehm E, Armbrust EV, Boessenkool KP (2008) Phytoplankton calcification in a high-CO_2 world. *Science* **320**, 336–340.

Inoue S, Kayanne H, Yamamoto S, Kurihara H (2013) Spatial community shift from hard to soft corals in acidified water. *Nature Climate Change* **3**, 683–687.

Jackson R (2019) Eunice Foote, John Tyndall and a question of priority. Notes and Records. doi:10.1098/rsnr.2018.0066.

Jury CP, Whitehead RF, Szmant AM (2010) Effects of variations in carbonate chemistry on the calcification rates of *Madracis auretenra* (= *Madracis mirabilis* sensu Wells, 1973): Bicarbonate concentrations best predict calcification rates. *Global Change Biology* **16**, doi 10.1111/j.1365-2486.2009.02057.x.

Keeling CD, Piper SC, Bacastow RB, Wahlen M, Whorf TP, Heimann M, Meijer HA (2005) Atmospheric CO_2 and $^{13}CO_2$ exchange with the terrestrial biosphere and oceans from 1978 to 2000: Observations and carbon cycle implications. In: Ehleringer J. R., Cerling T. E., Dearing M. D. (eds.), *A History of Atmospheric CO_2 and its effects on Plants, Animals, and Ecosystems*, Springer Verlag, New York, pp 83–113.

Keeling RE, Kortzinger A, Gruber N (2010) Ocean deoxygenation in a warming world. *Annual Review of Marine Science* **2**, 199–229.

Killian CE, Wilt FH (2008) Molecular aspects of biomineralization of the echinoderm endoskeleton. *Chemical Reviews* **108**, 4463–4474.

Kleypas JA, Buddemeier RW, Archer D, Gattuso J-P, Langdon C, Opdyke BN (1999) Geochemical consequences of increased atmospheric carbon dioxide on coral reefs. *Science* **284**, 118–120.

Koch M, Bowes G, Ross C, Zhang X-H (2013) Marine macro-autotrophs and climate change. *Global Change Biology* **19**, 103–132.

Kroeker KJ, Kordas RL, Crim RN, Singh GG (2010) Meta-analysis reveals negative yet variable effects of ocean acidification on marine organisms, *Ecology Letters* **13**, 1419–1434.

Kroeker KJ, Micheli F, Gambi MC, Martz TR (2011) Divergent ecosystem responses within a benthic marine community to ocean acidification. *Proceedings of the National Academy of Sciences USA* **108**, 14515–14520.

Kroeker KJ, Michell F, Bambi MC (2012) Ocean acidification causes ecosystem shifts via altered competitive interactions. *Nature Climate Change* **3**, 156–159.

Kroeker KJ, Kordas RL, Crim R, Hendriks IE, Ramajo L, Singh GS, Duarte CM, Gattuso J-P (2013) Impacts of ocean acidification on marine organisms: Quantifying sensitivities and interaction with warming. *Global Change Biology* **19**, 1884–1896.

Krumhardt KM, Lovenduski NS, Long MC, Levy M, Lindsay K, Moore, JK, Nissen C (2019) Coccolithophore growth and calcification in an acidified ocean: Insights from community earth system model simulations. *Journal of Advances in Modeling Earth Systems* **11**, 1418–1437. https://doi.org/10.1029/2018MS001483.

Kurihara H (2008) Effects of CO_2-driven ocean acidification on the early developmental stages of invertebrates. *Marine Ecology Progress Series* **373**, 275–284.

Kurihara H, Shirayama Y (2004) Effects of increased atmospheric CO_2 on sea urchin early development. *Marine Ecology Progress Series* **274**, 161–169.

Langer G, Geisen M, Baumann K-H, Klas J, Riebesell U, Thoms S, Young JR (2006) Species-specific responses of calcifying algae to changing seawater carbonate chemistry. *Geochemistry, Geophysics, Geosystems* **7**, Q09006, doi: 10.1029/2005GC001227.

Langer G, Nehrke G, Probert I, Ly J, Ziveri P (2009) Strain-specific responses of *Emiliania huxleyi* to changing seawater carbonate chemistry. *Biogeosciences* **6**, 2637–2646.

Lebrato M, Iglesias-Rodriguez D, Feely RA, Greeley D, Jones DOB, Suarez-Bosche N, Lampitt RS, Cartes JE, Green DRH, Alker B (2010) Global contribution of echinoderms to the marine carbon cycle: CaCO$_3$ budget and benthic compartments. *Ecological Monographs* **80**, 441–467.

Mackey KRM, Morris JF, Morel FMM, Kranz SA (2015) Response of photosynthesis to ocean acidification. *Oceanography* **28**, 74–91.

Marin F, Luquet G, Marie B, Medakovic D (2008) Molluscan shell proteins: Primary structure, origin and evolution. *Current Topics in Developmental Biology* **80**, 209–276.

Marubini F, Thake B (1999) Bicarbonate addition promotes coral growth. *Limnology and Oceanography* **44**, 716–720.

Morse JW, Andersson AJ, Mackenzie FT (2006) Initial responses of carbonate-rich shelf sediments to rising atmospheric pCO$_2$ and ocean acidification: Role of high Mg-calcites. *Geochimica et Cosmochimica Acta* **70**, 5814–5830.

Orr JC, Fabry VJ, Aumont O, Bopp L, Doney SC, Feely RA, Gnanadesikan A, Gruber N, Ishida A, Joos F, Key RM, Lindsay K, Maier-Reimer E, Matear R, Monfray P, Mouchet A, Najjar RG, Plattner G-K, Rodgers KB, Sabine CL, Sarmiento JL, Schlitzer R, Slater RD, Totterdell IJ, Weirig M-F, Yamanaka Y, Yool A (2005) Anthropogenic ocean acidification over the twenty-first century and its impact on calcifying organisms. *Nature* **437**, 681–686.

Parker LM, Ross PM, O'Connor WA, Borysko L, Raftos DA, Pörtner HO (2012) Adult exposure influences offspring response to ocean acidification in oysters. *Global Change Biology* **18**, 82–92.

Pauly D, Christensen V (1995) Primary production required to sustain global fisheries. *Nature* **374**, 255–257.

Pettit LR, Hart MB, Medina-Sánchez AN, Smart CW, Rodolfo-Metalpa R, Hall-Spencer JM, Prol-Ledesma RM (2013) Benthic foraminifera show some resilience to ocean acidification in the northern Gulf of California, Mexico. *Marine Pollution Bulletin* **73**, 452–462.

Politi Y, Arad T, Klein E, Weiner S, Addadi L (2004) Sea urchin spine calcite forms via a transient amorphous calcium carbonate phase. *Science* **306**, 1161–1164.

Prieto FJM, Millero FJ (2001) The values of pK$_1$ and pK$_2$ for the dissociation of carbonic acid in seawater. *Geochimica Cosmochimica Acta* **66**, 2529–2540.

Przeslawski R, Byrne M, Mellin C (2015) A review and meta-analysis of the effects of multiple abiotic stressors on marine embryos and larvae. *Global Change Biology* **21**, 2122–2140.

Rappuoli R, Mandl CW, Black S, De Gregorio E (2011) Vaccines for the twenty-first century society. *Nature Reviews Immunology* **11**, 865–872.

Raven JA, Smith FA (1980) Intracellular pH regulation in the giant-celled marine alga *Chaetomorpha darwinii*. *Journal of Experimental Botany* **31**, 1357–1369.

Reinfelder JF (2011) Carbon concentrating mechanisms in eukaryotic marine phytoplankton. *Annual Review Marine Science* **3**, 291–315.

Riddley M (2010) *The Rational Optimist*. Harper Collins Publishers, London, UK.

Riebesell U, Zondervan I, Rost B, Tortell PD, Zeebe RE, Morel FMM (2000) Reduced calcification of marine plankton in response to increased atmospheric CO_2. *Nature* **407**, 364–367.

Ries JB (2011) A physicochemical framework for interpreting the biological calcification response to CO_2-induced ocean acidification. *Geochimica et Cosmochimica Acta* **75**, 4053–4064.

Ries JB, Cohen AL, McCorkle DC (2009) Marine calcifiers exhibit mixed responses to CO_2-induced ocean acidification. *Geology* **37**, 1131–1134.

Rodolfo-Metalpa R, Houlbrèque F, Tambutté E, Boisson F, Baggini C, Patti FP, Jeffree R, Fine M, Foggo A, Gattuso J-P, Hall-Spencer JM (2011) Coral and mollusc resistance to ocean acidification adversely affected by warming. *Nature Climate Change* **1**, 308–312.

Ross PM, Parker L, O'Connor WA, Bailey EA (2011) The impact of ocean acidification on reproduction, early development and settlement of marine organisms. *Water* **3**, 1005–1030.

Rouco M, Branson O, Lebrato M, Iglesias-Rodríguez MD (2013) The effect of nitrate and phosphate availability on *Emiliania huxleyi* (NZEH) physiology under different CO_2 scenarios, *Frontiers of Microbiology* **4**, doi: 10.3389/fmicb.2013.00155.

Sabine CL, Feely RA, Gruber N, Key RM, Lee K, Bullister JL, Wanninkhof R, Wong CS, Wallace DWR, Tilbrook B, Millero FJ, Peng T-H, Kozyr A, Ono T, Rios AF (2004) The oceanic sink for anthropogenic CO_2. *Science* **305**, 367–371.

Shi D, Xu Y, Morel FMM (2009) Effects of the pH/pCO_2 control method on medium chemistry and phytoplankton growth. *Biogeosciences* **6**, 1199–1207.

Shi W, Zhao X, Han Y, Che Z, Chai X, Liu G (2016) Ocean acidification increases cadmium accumulation in marine bivalves: A potential threat to seafood safety. *Scientific Reports* **6**, 20197.

Sikes CS, Wilbur KM (1982) Functions of coccolith formation. *Limnology and Oceanography* **27**, 18–26.

Taylor AR, Chrachri A, Wheeler G, Goddard H, Brownlee C (2011) A voltage-gated H^+ channel underlying pH homeostasis in calcifying coccolithophores. *PLoS Biol* **9**(6), e1001085.

The Royal Society (2005) *Ocean Acidification Due to Increasing Atmospheric Carbon Dioxide*. Policy document 12/05 Royal Society, London. The Clyvedon Press Ltd., Cardiff.

Tunnicliffe V, Davies KTA, Butterfield DA, Embley RW, Rose JM, Chadwick Jr WW (2009) Survival of mussels in extremely acidic waters on a submarine volcano. *Nature Geoscience* **2**, 344–348.

Tyndall J (1859) Note on the transmission of radiant heat through gaseous bodies. *Proceedings of the Royal Society of London* **10**, 37–39.

Uthicke S, Ebert T, Liddy M, Johansson C, Fabricius KE, Lamare M (2016) Echinometra sea urchins acclimatized to elevated pCO_2 at volcanic vents outperform those under present-day pCO_2 conditions. *Global Change Biology* **22**, 2451–2461.

Valenzuela JJ, López García de Lomana A, Lee A, Armbrust EV, Orellana MV, Baliga NS (2018) Ocean acidification conditions increase resilience of marine diatoms. *Nature Communications* **9**, 2328. doi:10.1038/s41467-018-04742-3.

van Geen A, Takesue RK, Goddard J, Takahashi T, Barth JA, Smith RL (2000) Carbon and nutrient dynamics during coastal upwelling off Cape Blanco, Oregon. *Deep-Sea Research II* **47**, 975–1002.

Van Dover CL (2000) *The Ecology of Deep-Sea Hydrothermal Vents*. Princeton University Press, USA.

Venn AA, Tambutté E, Lotto S, Zoccola D, Allemand D, Tambutte S (2009) Imaging intracellular pH in a reef coral and symbiotic anemone. *Proceedings of the National Academy of Sciences* **106**, 16574–16579.

Walker CH, Livingstone DR (eds.) (1992) *Persistent Pollutants in Marine Ecosystems*. Pergamon Press, Oxford, p. 272.

Weiss R (1974) Carbon Dioxide in Water and Seawater: The solubility of a non-ideal gas. *Marine Chemistry* **2**, 203–215.

Wood HL, Spicer JI, Widdicombe S (2008) Ocean acidification may increase calcification rates, but at a cost. *Proceedings of the Royal Society London Series B* **275**, 1767–1773.

Young JR, Henriksen K (2003) Biomineralization within vesicles: The calcite of coccoliths. *Reviews in Mineralogy and Geochemistry* **54**, 189–215.

Zeebe RE, Wolf-Gladrow DA (2001) CO_2 *in Seawater: Equilibrium, Kinetics, Isotopes*. Elsevier Oceanography Series, Vol. 65, Elsevier, Amsterdam, p. 346.

Zondervan I, Zeebe RE, Rost B, Riebesell U (2001) Decreasing marine biogenic calcification: A negative feedback on rising atmospheric pCO_2. *Global Biogeochemical Cycles* **15**, 507–516.

Chapter 3

Ocean Warming

Continuing Increases in Greenhouse Gases from Human Activities Are Warming Earth

"*The highest effect of the sun's rays I have found to be in carbonic acid gas... An atmosphere of that gas would give to our Earth a high temperature; and if as some suppose, at one period of its history the air had mixed with it a larger proportion than at present, an increased temperature from its own action as well as from increased weight must have necessarily resulted.*"

— From "Circumstances affecting the heat of the Sun's rays" by Eunice Foote,[1] the first person to demonstrate the absorption of heat by water vapour and carbon dioxide, and hypothesize the connection between levels of carbon dioxide and the warming of Earth.

Introduction

The continued and fast increase in anthropogenic greenhouse gases as a direct result of human activities is the primary cause of a 0.65–1.06 °C rise in global average temperatures over the past few decades (IPCC, 2013, 2014; Reid, 2016). Specifically, global mean surface temperature has increased by ~0.85 °C between 1880 and 2012, and a further increase by 2.6–4.8 °C

---
[1] Foote E (1856) Circumstances affecting the heat of the Sun's rays. *American Journal of Science and Art* **22**, 382–383.

is anticipated under Representative Concentration Pathway (RCP) 8.5 (IPCC, 2013). Despite some agreement across nations to limit greenhouse gas emissions, global temperatures will continue to rise for the next few centuries given the inertia of the atmospheric and oceanic systems (IPCC, 2014).

The ocean plays a major role in the regulation of climate as it is the most important reservoir of heat, carbon and water on the planet, preventing the Earth's temperature from soaring (Reid *et al.*, 2009; Rhein *et al.*, 2013). Since the Industrial Revolution, the accumulation of heat near the Earth's surface has been causing global warming because the Earth's atmosphere is absorbing more energy from the sun than is radiated back into space (Rhein *et al.*, 2013; Trenberth *et al.*, 2014; von Schuckmann *et al.*, 2016), a phenomenon termed "Earth's energy imbalance" (EEI). A major concern with regard to the ocean uptake of excess energy gained by the Earth due to the EEI is how climate responds to radiative forcing.[2] The balance between the absorbed visible solar radiation and the outgoing longwave radiation determines the radiation budget of the Earth (Figure 1). The way greenhouse gases affect the Earth's radiation budget is by absorbing infrared radiation and releasing it back to the Earth's surface, thus increasing the Earth's energy balance and causing positive radiative forcing, ultimately leading to warming. On the other hand, negative radiative forcing results in a decrease in the energy budget and ultimately leads to cooling.

The accelerated warming of the Earth is of great concern to humanity because its negative impacts are far beyond the melting ice caps in Greenland and other polar regions, the rise in sea level, extreme weather conditions, the shrinking glaciers and the bleaching of corals. Indeed, temperature governs the nature and speed of metabolic pathways and the ways ecosystems are structured and function because warming impacts stratification, nutrient availability, and metabolic rates. Warming of the oceans could also affect the stability of gas hydrates, for example, methane,[3]

---

[2] *Radiative forcing* is defined as the difference between incoming solar radiation and outgoing infrared radiation caused by the increased concentration of a gas. It is expressed as W/m^2 (the rate of energy change, as a result of greenhouse gas emissions, per unit area of the globe).

[3] *Methane* (CH_4), like CO_2, is a greenhouse gas (although 84 times more potent than CO_2), i.e., it absorbs the sun's heat, warming the atmosphere. Methane is released in the production and transport of coal, natural gas, and oil as well as from livestock and other agricultural practices and by the decay of organic waste in landfills.

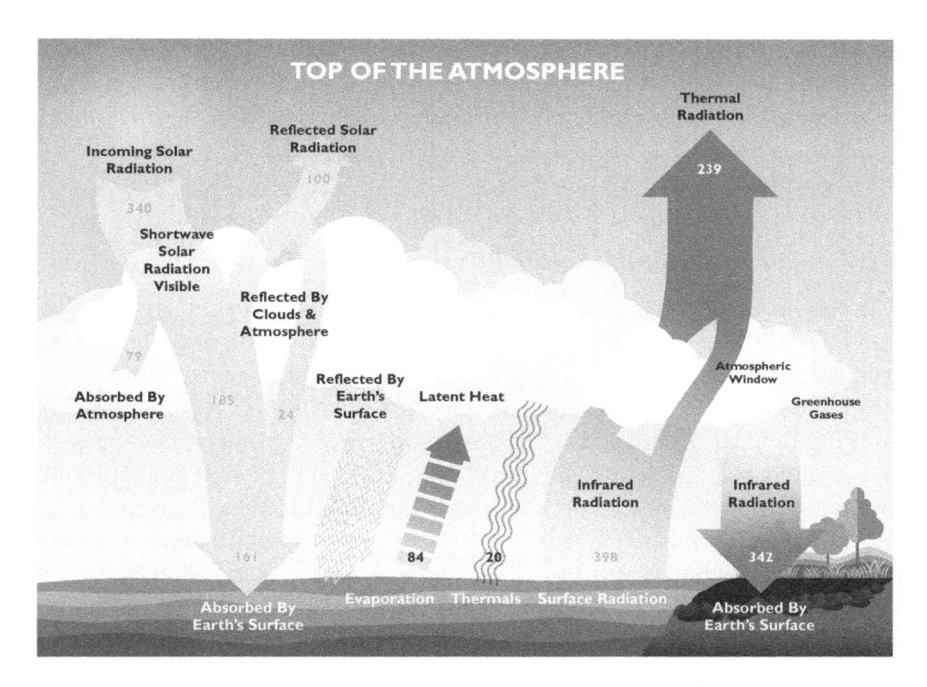

Figure 1. Representation of the global mean radiation balance of the Earth's climate conditions at the beginning of the 21st century (adapted from Wild *et al.*, 2013). Numbers (in W m^{-2}) indicate best approximations for globally averaged energy balance components (for uncertainty ranges, see Wild *et al.*, 2013).

a potent greenhouse gas. A concern is that if methane hydrate[4] deposits (locked beneath the ocean floor) are either warmed or depressurized they will revert to seawater and natural gas, which could exacerbate global warming. There is however no conclusive proof that hydrate-derived methane is being released to the atmosphere and more observations and improved numerical models are needed to forecast the climate–hydrate interactions in the future (Ruppel and Kessler, 2017).

This chapter explores the consequences of warming in marine ecosystems and the costs of marine ecosystem adaptation including repercussions

---

[4] *Methane hydrate* (also known as methane clathrate, hydromethane, methane ice, fire ice, and natural gas hydrate) is a crystalline solid that consists of a cage-like lattice of ice inside of which molecules of methane are trapped. It is found in sediments below the Arctic permafrost, along continental margins, in deep-water sediments of inland lakes and seas, and under Antarctic ice.

on human society. The chapter focuses on a number of marine model organisms and case studies to illustrate how warming impacts organism physiology and ecosystem function. Finally, one section discusses how shifts in phenology as well as organisms' physiology caused by ocean warming are leading to food insecurity in the oceans.

Challenges in Studying Global Temperature Trends

The consensus that humans are causing recent global warming is shared by 90–100% of the publishing climate scientists according to six independent studies (Cook *et al.*, 2016). We know that 13 of the warmest sea surface temperature (SST) years on record since 1880 (with the exception of 1997 and 1998) have occurred since 2000 (Reid, 2016). However, changes in global temperature can also be linked to variations in climate and episodic events, such as volcanic eruptions (e.g., Mt. Pinatubo eruption in 1991), which cause natural variability and can contribute to short-term warming or cooling (Trenberth and Dai, 2007). One example is El Niño Southern Oscillation[5] (ENSO) that can cause heat to come out of the ocean and be radiated to space (Trenberth *et al.*, 2002). In addition to the challenges in detecting ecosystem changes related to climate change, determination of temperature trends requires baseline data from which change can be evaluated. These data are however quite limited, especially from studies concerning ecosystem structure, functional properties, infectious disease epidemiology and behavior.

A challenge in studying warming trends to better inform our forecasts of future scenarios is to be able to determine whether observed changes in a given period are mostly the result of ongoing climate change or short-term climate variability. Specifically, when studying climate, one concern is whether the time frames under study are sufficiently long to determine a trend in climate change with any degree of certainty. Henson *et al.* (2010) proposed that 30–40 years are needed in most regions to distinguish between effects of climate change and climate variability in studying trends in phytoplankton productivity. Specifically, short-term data can

---

[5]The El Niño-Southern Oscillation (ENSO) is an oscillation of the large-scale ocean–atmosphere system in the tropical Pacific that causes warmer than average sea surface temperatures in the central and eastern equatorial Pacific as opposed to the cooler conditions associated with La Niña. El Niño has important consequences for weather conditions around the world.

detect regional and basin-scale temporal variability (e.g., ENSO, La Niña, Pacific Decadal Oscillation[6]) rather than long-term trends. For example, although the temperature of the Earth has been increasing since the start of industrialization (Abram et al., 2016), there have been warming hiatuses and fluctuations in land and ocean temperature. For example, the global land and sea surface temperature peaked in the 1940s and then remained stable until the 1970s (Hartmann et al., 2013; Dieng et al., 2017) (Figure 2).

Studies exploring trends over short timescales have found that between 1998 and 2013, the rate of warming in the surface of the Earth was modest compared to the warming in the second half of the 20th century (IPCC, 2013). This phase has been termed the "global warming hiatus" (IPCC, 2013; Roberts et al., 2015), which has been used by some climate skeptics as evidence for a slowdown in global warming (Reid, 2016). However, after this warming hiatus period, the rate of warming has been increasing and, for the first time since the beginning of established time series, record-breaking global temperatures were observed in both 2014 and 2015, suggesting that the so-called global warming hiatus (Hansen et al., 2015) was a short-term feature. Indeed, 2015 was identified as the warmest year within the 136 years of extended reconstructions of sea surface temperature records, and the fourth such record-breaking year since 2005 (NOAA National Centers for Environmental Information, 2016), and recently, 2017 was identified as the warmest year on record (Cheng and Zhu, 2018). The so-called warming hiatus has been interpreted by Yan et al. (2016) as energy redistribution within the oceans rather than a slowdown in warming of the climate system. These gaps are considered short-term features as both global surface temperatures including SSTs and their rates of change display a continued increasing and accelerating trend in the long term (Smith et al., 2015).

Since the 1970s, the oceans have been absorbing more than 90% of EEI caused by the greenhouse effect and other impacts of human activities, which have increased the ocean heat content as well as buffered temperature shifts in the atmosphere, land and sea surface (Wijffels et al., 2016). The residual heat (heat leftover) has been causing melting of land and sea ice, and warming of the atmosphere, land and ocean surface (Rhein et al., 2013;

---

[6]The Pacific Decadal Oscillation (PDO) is a long-lived El Niño-like pattern of Pacific climate variability that causes decadal-scale climate variations over the Pacific Ocean and, like ENSO, has important impacts on marine ecosystems and fisheries.

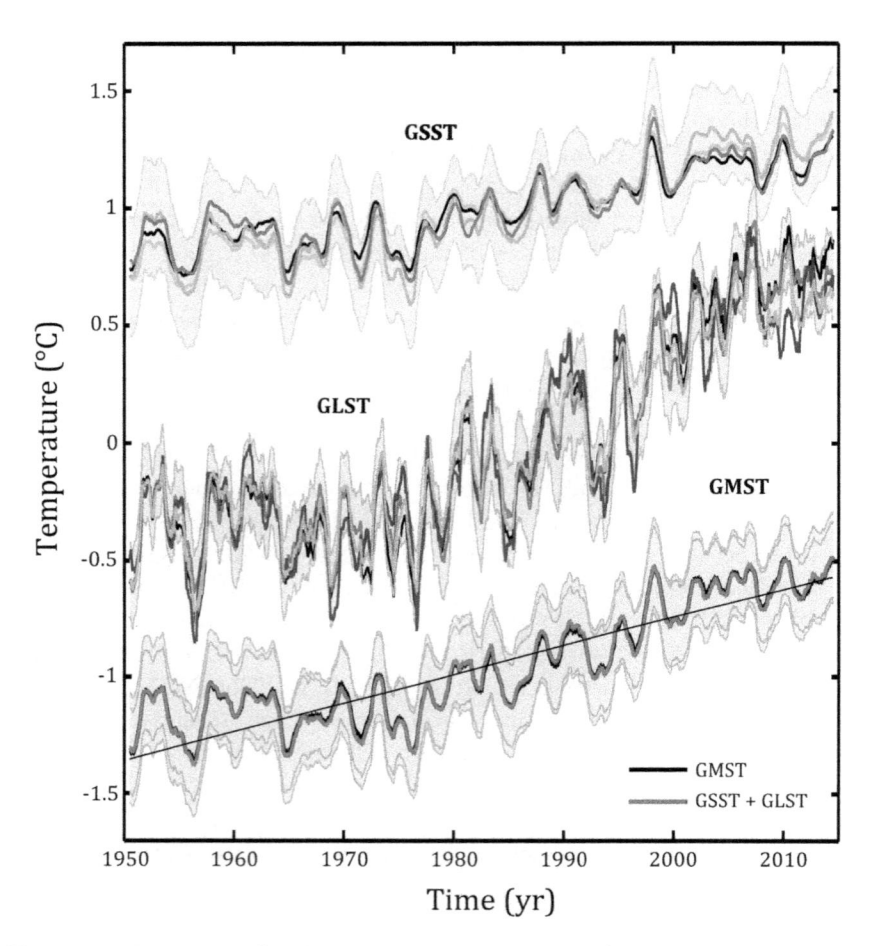

Figure 2. Time series for global temperature databases (from Dieng et al., 2017). Global sea surface temperature (GSST) time series from January 1950 to December 2014 averaged from four datasets: National Oceaonographic and Atmospheric Administration (NOAA)/Extended Reconstructed Sea Surface Temperature 4 (blue), Japan Meteorological Agency/COBE (black), Cowtan and Way (2014) (green) and associated uncertainty (shaded green) and Kaplan et al. (1998) (red). Global land surface temperature (GLST) time series from January 1950 to December 2014 derived from four processing groups: Global Historical Climatology Network and Climate Analysis and Monitoring System (GHCN_CAMS; black), Cowtan and Way (2014) (red) and associated uncertainty (shaded red), NASA (Goddard Institute for Space Studies (GISS); blue) and Berkeley (Temperature Average (TAVG); green). Global mean surface temperature (GMST) time series for January 1950 to December 2014 obtained through (1) average of NOAA_GHNC/Merged Land–Ocean Surface Temperature, Cowtan and Way (2014), NASA/GISS and Berkeley/TAVG (black) and associated uncertainty (light black line) and (2) sum of area-weighted "GSST plus GLST" (red) and associated uncertainty (shaded red) [see Dieng et al.(2017)] for details.

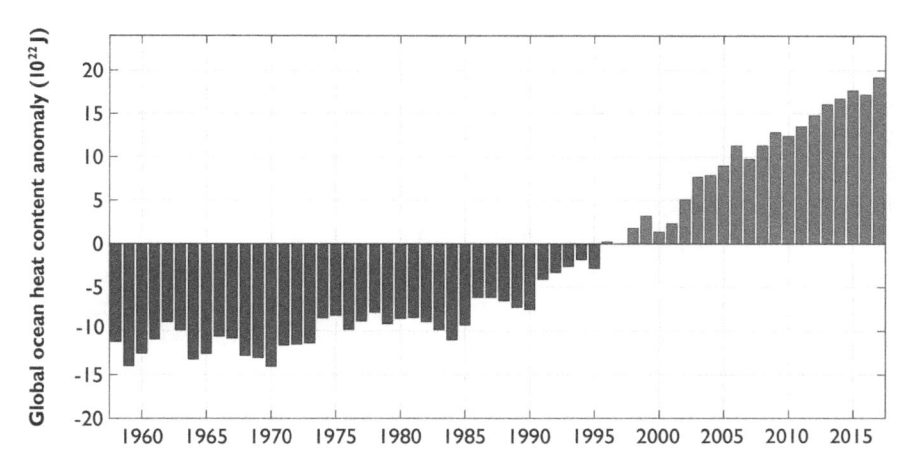

Figure 3. Change in global upper-level (0–2,000 m) ocean heat content since 1958. Each bar shows the annual mean relative to a 1981–2010 baseline. The final bar on the right shows the 2017 value. Reliable ocean temperature records date back to 1958.

Source: Institute of Atmospheric Physics, Chinese Academy of Sciences.

Wijffels *et al.*, 2016; Cheng *et al.*, 2017). The ocean heat content after 1980 increased fairly steadily and, since 1990, it has increasingly affected deeper layers of the ocean (Cheng *et al.*, 2017). According to a global ocean analysis by the Institute of Atmospheric Physics, Chinese Academy of Sciences (IAP, CAS; http://english.iap.cas.cn/RE/201801/t20180118_189348.html), 2017 was the warmest year on record, and the upper 2,000 m of the ocean were the warmest since 1958 (see also Cheng and Zhu, 2018) (Figure 3).

The Role of the Oceans in Regulating Global Warming

The major pieces of evidence supporting the scale of ocean warming were gathered in 2015 by the Grantham Institute to conclude that, if the same amount of heat that has gone into the top 2000 m of the ocean between 1955 and 2010 had instead accumulated into the lower 10 km of the atmosphere, then the Earth would have seen a warming of 36 °C (Whitmarsh *et al.*, 2015). Indeed, the oceans play a crucial role in counteracting the industrial-era warming, which started as early as the mid-19th century (Abram *et al.*, 2016). It has been estimated that more than 90% of the Earth's excess heat has been stored in the ocean in the last 50 years (Roemmich *et al.*, 2012) and, without the oceans, global temperatures would have risen to dangerous levels (Edwards, 2016). Specifically, because of its massive dimensions, and low albedo, the ocean can absorb massive amounts of heat. Seawater also has high density and specific heat,[7] enabling the oceans to store heat >1,000 times more efficiently than the atmosphere (Levitus et al., 2005).

One way the oceans respond to changes in the climate system's energy budget caused by warming is by thermal expansion, which causes sea-level rise (Levitus et al., 2005; Bindoff et al., 2007). As the ocean warms, the density of seawater decreases and therefore the volume of the ocean increases (Church et al., 2013). This thermal expansion (or steric sea-level rise) is one of the major contributors to sea-level rise during the 20th and 21st centuries. Indeed, the IPCC Fourth Assessment Report found that between 1961 and 2003, thermal expansion accounted for about 25% of the sea-level rise, melting of land ice accounted for less than 50%, and changes in land water storage accounted for less than 10% (Bindoff et al., 2007). According to Levitus et al. (2000), the world's oceans have stored 2×10^{23} Joules of heat between 1955 and 1995 with over half of this occurring in the upper 300 m for a warming rate of 0.7 °C/century. The steric sea-level rise equivalent is 0.55 mm/yr, with maxima in the subtropical gyre of the North Atlantic and the tropical eastern Pacific.

The extent of climate change is governed by the rate at which heat is removed from the ocean surface into the ocean interior such that if heat is taken up more readily, climate change is delayed, but sea level rises more rapidly. Although it is expected that sea-level rise will likely accelerate in the future because of global warming, there are key uncertainties including the possible role of the Greenland and West Antarctic ice sheets and the amplitude of regional changes that will determine the scale of sea-level rise (Nicholls and Cazenave, 2010).

The oceans are also a major sink for carbon dioxide (CO_2) and, as with other gases, the solubility of CO_2 in seawater decreases as temperature increases. The world's oceans currently take up ~25–30% of the anthropogenically produced CO_2 from the atmosphere, and the North Atlantic is a key component of this oceanic carbon sink (Sabine et al., 2004; Le Quere et al., 2015). In addition to being central to the uptake of CO_2, oceanic water masses redistribute heat, nutrients and carbon between the tropics and the poles through the Atlantic Meridional Oceanic

---

[7]The amount of heat per unit mass required to raise the temperature by 1 °C.

Circulation (AMOC)[8] (Ballantyne et al., 2012; Reid, 2016). A weakening of the AMOC is expected in the next century and it appears that ice melting has a long-term impact. Specifically, the AMOC is reduced by 47% when the melting of land ice is considered and represents an extreme melting scenario compared to ~21% without ice melting (Swingedouw et al., 2006). The main repercussions of any alterations of the AMOC will be impacts on global climate, ecosystems and biogeochemical cycles (Ganachaud and Wunsch, 2000; Schmittner, 2005). For example, as the oceans warm, their capacity to absorb the CO_2 that human activities are adding to the atmosphere diminishes. Similarly, warming of water depths distant from the air–water interface, and therefore not necessarily in equilibrium with the air, will also cause a decrease in the solubility of CO_2 and other gases such as oxygen. Whether or not and the extent to which marine biota will adapt fast enough to these environmental challenges caused by humans and how ecosystems will look like in the future remain open questions.

Climate model simulations suggest that significant oceanic and atmospheric warming is likely to continue unless current greenhouse gas emission levels are substantially reduced (Collins et al., 2013). We know that the warming of the oceans can cause shifts in biodiversity and ecosystem structure and function and that humans, particularly those living in the poorest regions, have already been negatively impacted. For example, the influence of warming on weather patterns is affecting productivity in coastal communities, risking food security and the preservation of valuable economies that depend on a healthy ocean. Assuming sustained economic and population growth and little move towards reducing fossil fuel consumption and emissions by key nations (particularly those with large economic power and population numbers), persistent warming will

---

[8] Commonly known as the "global conveyor belt", this global circulation system is governed by temperature and salinity (hence the term "thermohaline circulation"). The conveyor belt consists of near-surface warm northward water flow, compensated by a colder southward flow at depth accompanied by heat loss to the atmosphere at high latitudes in the North Atlantic. The dynamics of this circulation is driven by subpolar North Atlantic water, which is cold and salty and therefore denser, and sinks toward the ocean floor. Surface water moves in to replace the sinking water, thus creating a current between the continents, crossing the equator, and reaching Antarctica, where the water cools and sinks again, as it does in the North Atlantic (the conveyor belt gets reactivated). Around Antarctica, the current splits northward into the Indian and Pacific Oceans (see Clark et al., 2002).

place Earth in unknown territory in human history, with likely dangerous consequences.

How Do Marine Organisms Acclimate/Adapt to Warming?

One of the big challenges that ecosystems face is that the rate of warming driven by human activities is at least 10 times faster than any change in climate over the past 50 million years (Masson-Delmotte et al., 2013). All ocean basins have undergone significant warming since 1998 (Figure 4), and the greatest warming occurred in the southern oceans, the tropical/subtropical Pacific Ocean, and the tropical/subtropical Atlantic Ocean (Cheng et al., 2017). At present, SST may vary seasonally by 1–2 °C in tropical latitudes but over 10 °C in higher latitudes (Rummer and Munday, 2017). Therefore, tropical species are expected to be less resilient than their high-latitude counterparts because tropical species live in a more thermally stable environment. Indeed, tropical fish species appear to be less adaptable to warming that those from higher latitudes (see Wood and McDonald, 1997; Munday et al., 2012), and some reef species appear to be already close to their thermal limits for aerobic (Rummer et al., 2014), swimming (Johansen and Jones, 2011) as well as reproductive (Pankhurst and Munday, 2011) performances.

A consequence of warming that decreases metabolic performance is the fact that temperature governs O_2 solubility and uptake, transport and delivery in marine organisms (Pörtner, 2001; Pörtner and Knust, 2007; Pörtner and Farrell, 2008; Eliason et al., 2011). One parameter used to assess the oxygen available for processes other than basic functions, including reproduction, migration, and defense against predation, is aerobic scope, calculated as the difference in oxygen consumption between resting and maximal performance (Fry and Hart, 1948; Pörtner and Farrell, 2008). An organism will have the highest aerobic scope at a specific temperature beyond which the aerobic scope falls as oxygen supply to the tissues decreases (Gardiner et al., 2010; Rummer et al., 2014). When animals have to adjust fast to warming, their aerobic scope can decrease dramatically (e.g., by >70% in some reef fish living under 1–3 °C above their optimum) (Rummer et al., 2014). This could have major implications in their long-term adaptation (long-term evolutionary selection of individuals in the population) as metabolic trade-offs (e.g., less energy for reproduction if temperature increases exceed levels supporting the rate of adaptation) could result in a decline in ecological fitness.

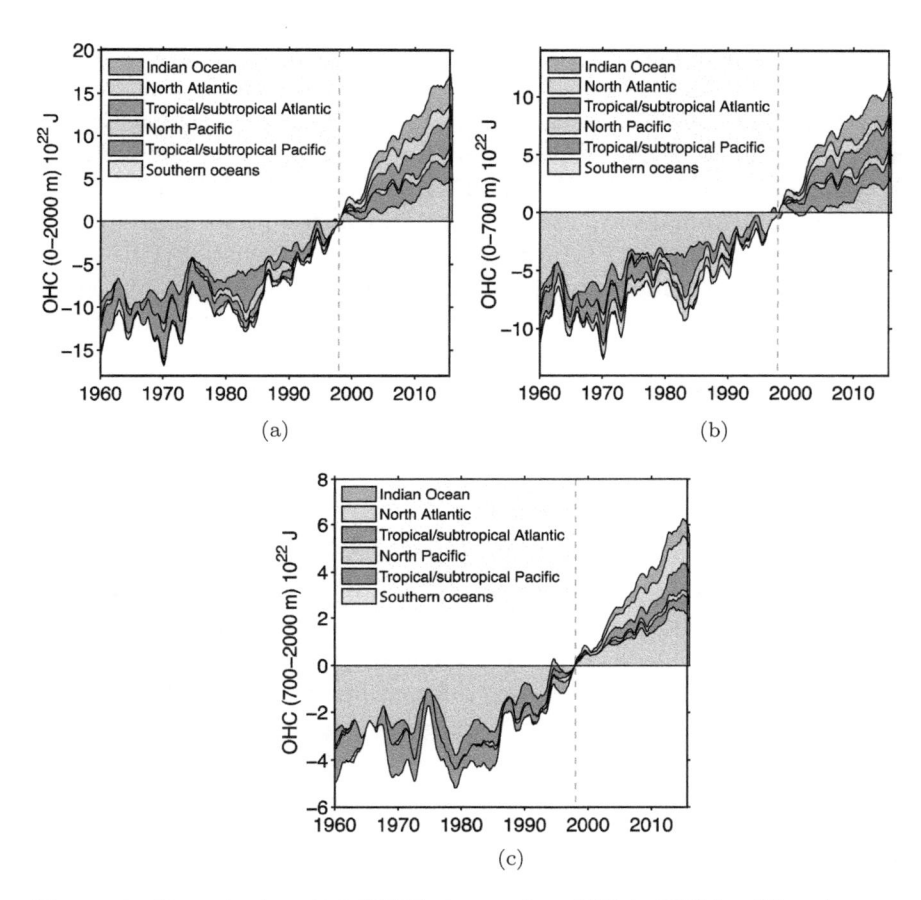

Figure 4. Ocean heat content (OHC) changes from 1960 to 2015 for different ocean basins (from Cheng et al., 2017): (a) 0–2,000 m; (b) 0–700 m; (c) 700–2,000 m. All the time series are relative to the 1997–1999 base period and smoothed by a 12-month running filter. The curves are additive, and the OHC changes in different ocean basins are shaded in different colors (see Cheng, 2017).

The energetic cost of acclimation (short-term responses to environmental change/within-generation plasticity) to warming can add to the costs associated with ocean acidification and other climate-driven stressors although the nature of the combined effects varies among organisms. On long timescales, transgenerational plasticity (plasticity across generations, i.e., phenotypic changes that persist for multiple generations) and heritable genetic traits allow some fish populations to be resilient and maintain performance as oceans warm (Donelson et al., 2012). For example, juveniles of the sheepshead minnow, *Cyprinodon variegatus*, grow faster at higher temperatures if their parents have been exposed to high-temperature conditions (Salinas and Munch, 2012). Another study exploring evolutionary adaptation to warming found plasticity and inheritable genetic variation in Chinook salmon (Munoz *et al.*, 2015). Because of this biological and environmental complexity, predicting the acclimation and adaptation of organisms to warming is not yet possible. However, what appears to be clear is that maintaining genetic variation will maximize the potential for adaptation to warming and climate change in general. Given that genetic heterogeneity is generally correlated with population size, increasing the extent of marine-protected areas and reducing fishing pressure are two approaches that are likely to aid in ameliorating the effects of environmental stressors including warming.

Poleward Migration of Species Driven by Warming

Some species will adapt to warming via phenotypic changes (acclimation) or through natural selection of individuals (adaptation), but all species have inherent limits in their capacity to respond to changing climate (Stillman, 2003; Williams *et al.*, 2008). Among the multiple ways by which organisms respond to warming are shifts in biogeography, contraction and expansion of species ranges, and alterations in species' abundances and activities (Figure 5) (Bates *et al.*, 2014; Fossheim *et al.*, 2015). There is some evidence indicating that the redistribution patterns of species can have propagating effects through entire ecosystems including changes in productivity and carbon storage. One example is the climate-driven poleward expansion of mangroves worldwide, at the expense of saltmarsh habitats, which is changing carbon sequestration (Cavanaugh *et al.*, 2014).

A consequence of species latitudinal redistribution is the change in diversity and composition of populations and communities and their functional traits within communities (Figure 5) (Buisson *et al.*, 2013; Pecl *et al.*, 2017). The distribution of species is already changing at accelerated rates both at regional and global scales (Pecl *et al.*, 2017), and it is expected that equatorial motile species will migrate to higher latitudes (Munday *et al.*, 2008; Nilsson *et al.*, 2009). The potential impacts of species temporal or spatial redistribution are changes in (a) the number of species coexisting at any given location (alpha diversity); (b) the community composition in space and time (beta diversity); and/or (c) the number of species found

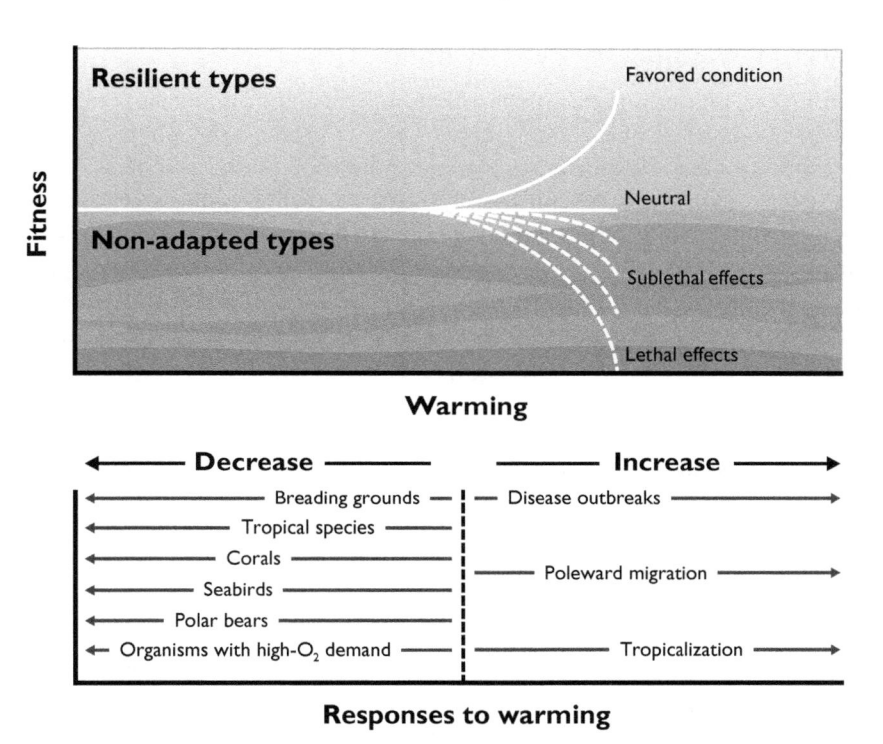

Figure 5. Elevated mean ocean temperatures may mean longer growth periods in some individuals that can profit from warming by increasing their reproduction (Eggert et al., 2005; Reusch, 2014). On the other hand, warming selects against individuals with a narrow range of tolerance to warming and/or those unable to migrate. For example, summer heat waves may cause sublethal stress in many marine groups including seagrasses (Reusch et al., 2005) and corals (Howells et al., 2011). Bottom: some responses to warming highlighting increasing (right) and decreasing (left) processes and organism abundance and health.

within a larger spatial range (gamma diversity) (Ochoa-Ochoa et al., 2012). Given that species diversity is considered to underlie functional diversity, any changes in the distribution patterns of species as the oceans get warmer will likely affect ecosystem function (Oliver et al., 2015; Pecl et al., 2017). On top of this, photosynthesis and other physiological processes such as protein stability, enzyme activity and membrane permeability (Kordas et al., 2011) can be altered in different species/groups as increasing temperature reaches levels causing stress or exceeding the organism's tolerance threshold.

Effect of Warming on Photosynthetic Organisms

Every year, photosynthetic carbon fixation by marine organisms is responsible for about 50 Gt of carbon, largely (~45 Gt) produced by phytoplankton, of which 16 Gt are exported to the ocean interior (Falkowski et al., 1998). Therefore, photosynthesis ultimately leads to carbon sequestration in the deep ocean, making phytoplankton important players in the carbon cycle on planetary scales. Warming can affect primary production directly, such as by increasing metabolic rates in cool regions that become warmer and by selecting specific groups of photosynthetic organisms that can survive in warmer scenarios. Additionally, rising temperature can have indirect detrimental effects on photosynthesis via upper-ocean warming-enhanced vertical stratification, which limits nutrient flux from deep waters towards the upper ocean, and can affect phytoplankton population composition and size structure, phenology (timing of seasonal processes), productivity, biochemical composition, and biogeography (Lewandowska et al., 2014; Mojica et al., 2015).

Although there is extensive research on the potential effects of warming (and acidification) on marine ecosystems, there is still considerable uncertainty regarding how biota will adapt and how ecosystem services (provided by organisms' processes such as photosynthesis in phytoplankton and macroalgae, calcification by corals and plankton, nitrogen fixation in photosynthetic bacteria) will be altered under future climate scenarios. One parameter that has been used to measure the rate of change of a biological process as a function of temperature is the unitless temperature coefficient Q_{10}, which can be calculated as follows:

$$Q_{10} = [R_2/R_1]^{[10/(T_2-T_1)]}$$

where Q_{10} represents the factor by which the rate (R) of a reaction increases for every 10 °C rise in temperature (T), where R_1 and R_2 are the measured reaction rates at temperatures T_1 and T_2 (in °C), respectively (where $T_1 < T_2$).

For example, Q_{10} can explain the selection of processes under ocean warming scenarios, such as the apparent dominance of heterotrophic over autotrophic processes (López-Urrutia et al., 2006; Sommer and Lengfellner, 2008). Specifically, the Q_{10} value of light-saturated phytoplankton growth (1.88; Eppley, 1972) is close to the lower end of Q_{10} values reported for heterotrophic processes (1.8–5.2), for example, in bacteria (Sand-Jensen et al.,

2007), phytoplankton (Hancke and Glud, 2004), and zooplankton (Ikeda et al., 2001).

It is still unclear whether rising temperatures cause a decline or an increase in phytoplankton biomass and productivity. Some studies suggest a decrease (Behrenfeld et al., 2006; Boyce et al., 2010; Sommer et al., 2012), whereas others propose an increase in net primary production as a result of warming and also because of changes in light, nutrients and grazing (Laufkötter et al., 2015). Warming can also impact the cell size composition of phytoplankton populations. Specifically, shrinking body size has been proposed as a universal response to warming (e.g., Falkowski and Oliver, 2007; Moran et al., 2010), although there are other factors driven by warming, such as nutrient limitation or shifts in the size-selective grazing that can result in smaller cell sizes (Higini and Sommer, 2013). The fate of organic carbon (the carbon content of cells other than carbon biomineral, such as carbonate, produced by some cells) in the ocean's water column appears to be dependent on phytoplankton size, and a shift towards smaller phytoplanktonic species could arguably result in decreased export flux (Passow and Carlson, 2012). However, some studies have suggested that small cells can also be aggregated into larger sinking masses in proportion to their abundance (Richardson and Jackson, 2007). Although still under debate, a decrease in phytoplankton cell size may be an important consideration when forecasting the flux of carbon to the deep sea floor, which is the most important way by which the Earth "locks" carbon away from the atmosphere.

Effect of Warming on Coral Ecosystems

Although coral reefs occupy less than 0.1% of the ocean sea floor surface area, they provide a habitat for more than a quarter of all marine fish species and represent almost $9.8 trillion globally each year as social, economic and cultural services (Costanza et al., 2014). In addition to other human-driven stressors such as eutrophication and ocean acidification, ocean warming is a major factor governing coral health, causing chronic as well as abrupt impacts during heat waves. A phenomenon associated with warming is "coral bleaching", often driven by thermal stress during marine heat waves, which causes the disruption of the symbiotic relationship between corals and their algal symbionts, resulting in the algae leaving the coral and hence leaving the white skeleton visible through the transparent coral tissue. As a result, coral bleaching causes a decrease in coral growth and reproduction, increased susceptibility to disease and, in some cases, coral death (Baker et al., 2008).

The rate of decadal warming in areas dominated by coral reefs has increased by five-fold from the last century to the recent decades (1985–2012) (Heron et al., 2016a) and, over the last few decades, prolonged warming has caused bleaching to occur more and more frequently (Heron et al., 2016b). This increasing warming trend has been causing devastating effects on coral ecosystems and the humans depending on them and, since the 1980s, warming events have triggered unprecedented mass bleaching of corals, particularly three pan-tropical bleaching events in 1998, 2010 and 2015–2016 (Heron et al., 2016b). The severity and geographic extent of bleaching appears to be explained by the magnitude and spatial distribution of sea surface temperature anomalies (deviations from the mean) (Hughes et al., 2017). These events can be extremely disruptive in the long term as reef recovery can take 10–15 years in coral species that are fast colonizers (Kayanne et al., 2002; Gilmour et al., 2013), and coral replacement after coral death can takes many decades (Hughes et al., 2017). Analysis used in the IPCC Fifth Assessment Report predicted that the current warming trajectory will cause annual coral bleaching for almost all reefs and loss of corals from most coral ecosystems globally by 2050 (Hoegh-Guldberg et al., 2014).

Among modeling methods to assess the impacts of bleaching is the degree heating months (DHM), calculated as one month of SST that exceeds the 1985–2000 maximum monthly mean (Donner et al., 2005). It has been suggested that the increase in thermal tolerance required for reefs to adapt to warming is 0.2–0.3 °C per decade (Donner et al., 2005) and the extent to which symbionts can expand their thermal tolerance via shifts in their community composition may be limited (Hoegh-Guldberg et al., 2002). The conclusion from these modeling studies was that the frequency of coral reef bleaching globally could increase to being an annual or biannual event in 30–50 years due to climate change without an increase in the thermal tolerance of corals and their symbionts (Donner et al., 2005).

The effects of global warming on corals appear to be exacerbated when additional stressors such as ocean acidification operate simultaneously (Hoegh-Guldberg et al., 2007; Rodolfo-Metalpa et al., 2011; Prada et al., 2017). For example, Eastern Tropical Pacific reefs display resilience to acidification or warming when these stressors act alone, but when combined, the results can be devastating. For example, in the Southern Galapagos Islands, the combination of acidification and warming during the decade following the 1982-83 El Niño caused the reef framework to be completely eliminated (Manzello et al., 2017). It has been suggested that temperature overrides the effects on survival (Findlay et al., 2010) and ocean acidification largely affects calcification (Kroeker et al., 2013). Similarly, conditions that amplify warming, such as weak wind conditions that decrease water flow during heat wave events, can lead to mass coral bleaching and coral death (e.g., DeCarlo et al., 2017).

Effect of Warming on Fish Populations

In addition to the negative impacts of other anthropogenic stressors (e.g., acidification, deoxygenation, pollution, fishing pressure), warming alters the ecological fitness of fish populations and the components of the food web that fish rely on. Warming can shift the distribution of some fish populations polewards and to deeper waters, compress their habitat as waters get depleted in oxygen, and alter their phenology and physiology (Pinsky et al., 2013; Montero-Serra et al., 2015; Punzon et al., 2016; Checkley et al., 2017). Other ways for fish populations to survive is via acclimation, adaptation, and also via transgenerational adaptation (the capacity to adapt to warming over generations[9]). The latter has been observed in reef fishes (Donelson et al., 2012; Donelson and Munday, 2015; Munday et al., 2017), although the nature of adaptation mechanisms to warming in the long term remains unknown.

The majority of fish are ectotherms, meaning that they do not produce heat to maintain a constant body temperature (as is the case for birds and mammals), and in these animals, body temperature is a direct function of water temperature. In almost all ectotherms, any increase in temperature causes an increase in basal energy and thus oxygen requirements (Pörtner, 2001). On the other hand, endothermic fish such as tuna are unusual in that they have warm muscles through a well-developed vascular heat exchange system and high levels of myoglobin in red muscle (Stevens and Carey, 1981). This design allows them to maintain internal temperatures of up to 20 °C above ambient, an evolutionary advantage to increase the delivery of oxygen to the mitochondria by myoglobin (Stevens and Carey, 1981). Indeed, tuna have far higher rates of oxygen uptake than other fish and close to those of mammals, perhaps linked to their extremely high swimming speeds making them able to migrate across ocean basins in just a few months (Mather et al., 1995; Block et al., 2005).

---

[9]Transgenerational adaptation refers to the process by which parental environments shape offspring phenotypes.

A big concern is the extent to which tuna will be able to adapt to warming, particularly in light of other stressors such as overfishing, which are known to decrease population size, genetic diversity, and resilience to environmental change (Conover and Munch, 2002; Berkeley et al., 2004; Porch and Lauretta, 2016; Ward et al., 2016). For example, warming appears to induce metabolic stress more quickly in bluefin tunas because they have higher oxygen demands than warm-water *Thunnus* species (Block et al., 2005; Kitagawa et al., 2006). Additionally, climate change including shifts in temperature-mediated oxygen availability could affect spawning areas, which would require adaptation of adults and larvae and/or changes in migration patterns to target new spawning grounds (Block et al., 2005; Muhling et al., 2017). It appears that increasing temperatures may alter the spatial distribution of tuna populations, and it has been predicted that warming and other stressors, such as declining zooplankton densities (prey for tuna) may lower the carrying capacity for tuna and other commercially valuable fish by 2–5% per decade over the 21st century (Woodworth-Jefcoats et al., 2017).

Ocean warming also causes generally a decrease in the body size of fish (Daufresne et al., 2009; Gardner et al., 2011; Ohlberger, 2013), with likely consequences on community dynamics and ecosystem function and productivity (Millien et al., 2006; Vindenes et al., 2014). In ectothermic fish, the supply of oxygen has been proposed as the main cause of the temperature–size rule (Atkinson et al., 2006; Forster et al., 2012), where oxygen supply may limit maximum attainable body size (Czarnoleski et al., 2015). These reductions in maximum body size are stronger in larger species (Daufresne et al., 2009; Forster et al., 2012) and in tropical and polar environments, as these often have evolved in more stable thermal conditions (Tewksbury et al., 2008; Somero, 2010). For example, coral reef fishes are particularly sensitive to warming because, unlike in mid and high latitudes, temperature fluctuations in these environments are modest, and temperatures of just 1.5–3 °C above present-day summer averages cause declines in fitness (e.g., decreased aerobic scope, growth, reproduction, swimming ability) (Munday et al., 2008; Johansen et al., 2014). Many fish migrate from tropical to higher latitudes and this type of migration is expected to have impacts on several trophic levels. For example, a "tropicalization" of fish populations (the increase in the proportion of warm water species) with ocean warming is already increasing fish herbivory and hence exerting pressure on kelp forests (Verges et al., 2016).

Effect of Warming on Marine Birds

Marine top predators such as birds often feed over large oceanic areas and are exposed to varying environmental conditions. Warming and many other anthropogenic impacts (e.g., release of heavy metals, organochlorides, hydrocarbons and plastics into the oceans) are major threats to seabirds globally. Another issue affecting seabirds is overfishing because most seabirds are piscivorous, and two-thirds of the world's fish stocks are overexploited by industrial fisheries, which can lead to starvation in many seabird populations.

Results from some studies suggest that warming is expected to reduce phytoplankton and zooplankton biomass (Chust et al., 2014) and thus decrease the availability of prey for seabirds and, as a result, warming is affecting the phenology of seabirds, although the susceptibility to warming and other climate stressors depends on foraging methods, wing morphologies, and diving abilities. However, there are exceptions when warming can be beneficial for seabirds. For example, in the Bering Sea, transitions from cold to warm regimes may enhance productivity of piscivorous seabirds when the young of large predatory species of fish are numerous enough to provide forage (Hunt et al., 2002). Several observations of short-term responses of seabirds to climate change have revealed a detrimental impact on breeding success and body condition largely as a result of the need for seabirds to travel longer distances to find optimal feeding grounds, which increases energy expenditure (Dorresteijn et al., 2012, Péron et al., 2012).

In tropical regions, where many species are threatened by climate change, impacts of warming are often seen during El Niño years, (Gaston, 2001). For example, warming during El Niño can influence the foraging success of tropical seabird species and cause massive chick mortality and have devastating consequences on population size (Erwin and Congdon, 2007).

In the Southern Ocean, penguins and other seabirds that depend on krill as food are being negatively affected by climate change because warming is impacting krill (*Euphausia superba*) recruitment by reducing the area of sea ice formed near the Antarctic Peninsula (Loeb et al., 1997). In these high latitudes, breeding attempts by king penguins (*Aptenodytes patagonicus*) (Petry et al., 2013) and macaroni penguins (*Eudyptes chrysolophus*) (Gorman et al., 2010) have been observed at new sites. Similarly, long-term tracking data of king penguin breeding on the Crozet Islands (southern Indian Ocean) indicate that SST was the main driver controlling penguins' foraging distribution (Peron et al., 2012). Projections of the IPCC suggest a southward shift of foraging zones, ranging from 25 to 40 km per decade depending on the IPCC scenario. This will result in a doubling of traveling distances by 2100 in order for incubating and brooding birds to reach optimal foraging zones. This scenario is likely to be detrimental to the Crozet population in the long term, unless penguins develop alternative foraging strategies (Peron et al., 2012).

Effect of Warming on Marine Mammals

Temperature determines the distribution of marine mammals. Like pursuit-diving seabirds (e.g., penguins, some cormorants), pinnipeds (seals and sea lions) are widespread in ocean waters with summer SSTs below 20 °C while they are uncommon in warmer latitudes (Cairns et al., 2008). Like terrestrial animals, individual and population fitness are critically dependent on the synchronicity between fundamental life functions and seasonal availability of resources (Paul et al., 2008). Most alterations in marine mammal population numbers and health as a result of warming are and will continue to be a consequence of changes in the supporting food web driven by shifts in temperature, leading to altered biological productivity and food availability. For example, warming in the Southern Ocean is likely to result in the distribution of prey for southern elephant seals shifting either polewards and/or to increasing depths (McIntyre et al., 2011). Indeed, as a result of warming, seals dive to deeper waters in search for food and spend less time at targeted dive depths, possibly making them less successful foragers in environments undergoing warming, such as the Southern Ocean (McIntyre et al., 2011).

In the Arctic, polar bears (*Ursus maritimus*) are particularly vulnerable to warming because it is causing loss of sea ice that bears depend upon to hunt seals for a sufficient amount of time each year, to accumulate fat and survive periods without seals. Specifically, progressively earlier breakup of ice in spring is providing less time to access prey, for example, young seals. A study measuring simultaneously field metabolic rates, daily activity patterns, body condition, and foraging success of polar bears moving on the spring sea ice revealed that metabolic rates were 1.6 times greater than previously assumed. The authors argue that this result coupled with low intake of fat-rich marine mammal prey and high metabolic demands required for walking on thin ice led to an energy deficiency for more than half of the bears examined (Pagano et al., 2018). The immediate consequences are longer periods without food, poorer body condition, fewer and smaller cubs, and lower survival of cubs and old bears. In the long term, polar bears might remain confined to the Northern portions of the Arctic and become extinct in the Southern Arctic region by the middle of the century (Stirling and Derocher, 2012). Bears also change their behavior as habitat conditions change, presumably in an attempt to balance prey availability with energy costs (Ware et al., 2017).

Effect of Warming on Food Security

In addition to loss of species with narrow thermotolerance and poleward migrations of organisms among many other consequences, global warming also impacts marine food security (Table 1). Sources of marine protein such as fishes and invertebrates generally shift their distribution in response to ocean warming to higher latitudes and deeper waters, and it has been suggested that fisheries is likely to be affected by tropicalization of catch (Perry et al., 2005; Dulvy et al., 2008; Cheung et al., 2012). Deep-sea environments are also considered at risk from climate change (Sweetman et al., 2017) and many studies have already revealed impacts of warming (Purkey and Johnson, 2010) as well as other climate-driven phenomena, including deoxygenation (Stramma et al., 2012; Keeling et al., 2010; Helm et al., 2011), acidification of intermediate deep waters (Byrne et al., 2010), and decreased flux of organic matter to the seafloor (Ruhl and Smith, 2004; Smith et al., 2013).

In addition to warming of the oceans by 3 °C (2.1–4.0 °C) by the end of the century (Collins et al., 2013), extreme episodic phenomena, such as heat waves, are predicted to increase in frequency and magnitude as a consequence of global warming (Jentsch et al., 2007; Hegerl et al., 2011; Wernberg et al., 2013). Heat waves can cause tropicalization of ecosystems, decrease their diversity (Sorte et al., 2010; Wernberg et al., 2013) and turn ecosystems into a depauperate state (Wernberg et al., 2013; Verges et al., 2016). These extreme events, which can last for a few months, represent snapshots of what future scenarios could look like. Warming trends are also shaped by regular phenomena such as ENSO events, leading to warming anomalies, which appear to have increased in intensity and frequency as a result of climate change (Cai et al., 2014).

Warming will have an effect on the phenology of some marine animals and plants including fish recruitment as well as higher/lower trophic levels interacting with these populations. For example, a study assessing the

Table 1. Some effects of global warming on ocean processes and their impact on abiotic and biotic responses.

Processes affected by warming	Observed/anticipated consequences	
	Abiotic	Biotic
Ocean heat content	Rising temperatures at surface and depth[a] Shallowing of the pycnocline Intensification of ENSO events rising sea level[b]	Poleward migration physiological shifts coral bleaching higher disease incidence
Currents and heat transport	Increased poleward heat transport flooding from storm surges	Altered connectivity and dispersal of populations
Stratification	Increased stratification	Collapses in fisheries human malnutrition
Sea level (from warming and melting ice sheets and glaciers)	Rising sea level loss of islands, land inundation coastal erosion salt water intrusion into aquifers	Loss of some coral atolls human migration
Warming of adjacent land masses	Melting permafrost shrinking of mountain glaciers Surface melting of Greenland ice sheet	Shifts in terrestrial flora increased extent and magnitude of forest fires
Cryosphere melting	Thinning of ice shelves in Antarctica/increase of Antarctic ice mass and sea ice shrinking of Greenland ice sheet shrinking of sea ice in the Arctic and the Antarctic Bellingshausen/Amundsen seas	Decline in polar bear population fitness in the Arctic
Methane gas hydrate dissociation	Release of methane to the atmosphere from seafloor and permafrost	
Hydrological cycle	Increased poleward atmospheric moisture and precipitation extreme droughts and floods Rising salinity in mid latitudes and lower salinity in rainy tropical and polar seas	
Ocean carbon sink	Decreased oceanic CO_2 uptake from the atmosphere enhanced ocean acidification	Proliferation of non-calcifying organisms

(*Continued*)

Table 1. (*Continued*)

Processes affected by warming	Observed/anticipated consequences	
	Abiotic	Biotic
Deoxygenation	Reduced O_2 solubility and penetration into deeper water due to enhanced stratification	Stress and possible metabolic trade-offs in marine animals
Sulphur cycle	Possible decrease of oceanic sulphur release from the ocean to the atmosphere	Reduced production of dimethylsulphide (DMS) by phytoplankton
Biological carbon pump	Potential slowdown of the biological carbon pump	
ENSO and weather events	Increase in ENSO frequency/magnitude increase in frequency and magnitude of cyclones, hurricanes, monsoons, forest fires	Selection for organisms adapted to abrupt thermal shifts

[a]There is, however, some evidence of deep cooling (see Liang *et al.*, 2015, *Journal of Climate*).
[b]Water expands as it warms, resulting in sea-level rise.

phenology of fish species harboring the most productive upwelling systems (those in the eastern boundaries of continents) revealed that 39% of phenological events occurred earlier in recent decades and that zooplankton did not shift their phenology synchronously to conclude that these mismatches could potentially decrease fish recruitment (Asch, 2015). Warming also causes biological regime shifts, such as the shift in the mid-1970s from an "anchovy regime" dominating under cool conditions to a warm "sardine regime" in the Pacific Ocean (Chavez *et al.*, 2003).

Although there is plentiful evidence of climate-related decreases in marine species diversity and abundance, there are also examples of marine ecosystems showing notable resilience against extreme climatic events (O'Leary *et al.*, 2017). Marine organisms can acclimate and adapt to progressive and chronic warming by migrating to higher latitudes (in organisms that are mobile or have a pelagic life-cycle stage), adjusting their physiology to survive and maintain their ecological standing, acquiring/losing functions, changing their reproductive success (fitness), undergoing evolutionary selection of phenotypes/genotypes, and through developmental and transgenerational plasticity. Polar ecosystems are likely to change in structure given that, since the 1990s, the atmosphere in polar regions has been warming twice as fast as the average rate of global warming. Additionally, reduced sea ice extent will increase human access to polar fisheries. Tropical ecosystems are also predicted to be particularly susceptible to ocean warming because the tropics are relatively stable thermal habitats, and tropical animals and plants generally have a narrower range of thermal tolerance than temperate organisms (Tewksbury et al., 2008; Sunday et al., 2011). These are general trends although there are examples of many reef fishes that have large latitudinal distributions, spanning temperature ranges predicted for global warming scenarios (Feary et al., 2010; Burt et al., 2011), suggesting that some species might be able to adapt.

Some Solutions

- Reduce CO_2 emissions to ameliorate warming. This will be a long-term solution, but major cuts in CO_2 emissions need to be implemented urgently.
- Help ecosystems to adapt. For example, cut down pollution from rivers and coastal human developments through urban and industrial activities, which cause water eutrophication. Eliminating or reducing stressors additional to warming will help reduce stress in organisms and increase the probability of adaptation.
- Implement seasonal management tactics (e.g., fishery closures) to counteract the effect of mismatch between zooplankton and fish phenology caused by warming and other climate stressors.
- Bring incentives to communities to shift social behavior to a more sustainable economy by reducing consumption and production, and offering alternative choices to consumers.
- Promote a responsible industry that ensures sustainable production as well as processing of the end product. For example, after products and their packaging leave warehouses, most producers are not responsible for their fate. However, there are a few exceptions such as carbonated drinks that are redeemed for deposit, and rechargeable batteries and mercury thermostats that are recycled through manufacturer-sponsored programs.
- Vote for politicians that provide industry and municipalities with incentives and the power to deal with waste in an environmentally responsible way.

- Implement technology that can reduce unnecessary waste, remove waste and clean the environment.

Key Points

- Human activities are the primary cause of global warming.
- The oceans play a major role in the regulation of climate, preventing the Earth's temperature from spiraling.
- Warming causes melting of ice caps in polar regions, sea-level rise, extreme weather, shrinking of glaciers, bleaching of corals, stratification, nutrient limitation, and changes in metabolic rates among other impacts.
- A warming hiatus identified between 1998 and 2013 was a short-term feature and the oceans continue to undergo warming.
- Some negative consequences of warming in marine life include decreased metabolic performance, metabolic trade-offs (e.g., less energy for reproduction), a need to migrate to higher cooler latitudes.
- Recent warming events have triggered unprecedented mass bleaching of corals.
- Current warming trends will cause coral bleaching for almost all reefs and loss of corals from most coral ecosystems globally by 2050 (Hoegh-Guldberg et al., 2014).
- It appears that increasing temperatures will likely alter the spatial distribution of fish including commercial groups such as tuna.
- Warming generally causes a decrease in the body size of fish.
- Warming decreases the breeding success and body condition of marine birds largely because seabirds need to travel longer distances to find food, which increases energy expenditure.
- Warming is likely to affect the phenology of many marine organisms.
- Heat waves can cause tropicalization of ecosystems, decrease their diversity, and deteriorate ecosystems.

Questions

1. Define Earth's energy imbalance.
2. Why is Earth getting warmer?
3. What are the main causes of sea-level rise?
4. See how to calculate sea-level rise: http://www.antarcticglaciers.org/glaciers-and-climate/estimating-glacier-contribution-to-sea-level-rise/.
5. What is the relationship between temperature and CO_2 in seawater?

6. If $Q_{10} = 2$, how many times faster would an enzymatic reaction take place at 25 °C compared to 5 °C?
7. What condition would be more stressful and potentially lethal for an ectothermic fish: extremely high temperatures (38 °C) or temperatures between 5 °C and 10 °C?
8. How does O_2 metabolism change with increasing seawater temperature in marine animals?
9. How does warming affect marine mammals and birds?
10. How does bleaching affect the health of corals as a result of heat waves?

References

Abram NJ, McGregor HV, Tierney JE, Evans MN, McKay NP, Kaufman DS, the PAGES 2k Consortium (2016) Early onset of industrial-era warming across the oceans and continents. *Nature* **536**, 411–418.

Asch RG (2015) Climate change and decadal shifts in the phenology of larval fishes in the California Current ecosystem. *Proceedings of the National Academy of Sciences of the United States of America* **112**, E4065-E4074.

Atkinson D, Morley SA, Hughes RN (2006) From cells to colonies: At what levels of body organization does the 'temperature-size rule' apply? *Evolution and Development* **8**, 202–214.

Baker AC, Glynn PW, Riegl B (2008) Climate change and coral reef bleaching: An ecological assessment of long-term impacts, recovery trends and future outlook. *Estuar. Coast. Shelf Sci.* **80**, 435–471.

Ballantyne AP, Alden CB, Miller JB, Tans PP, White JWC (2012) Increase in observed net carbon dioxide uptake by land and oceans during the past 50 years. *Nature* **488**, 70–73.

Bates AE, Pecl GT, Frusher S, Hobday AJ, Wernberg T, Smale DA, Sunday JM, Hill NA, Dulvy NK, Colwell RK, Holbrook NJ, Fulton EA, Slawinski D, Feng M, Edgar GJ, Radford BT, Thompson PA, Watson RA (2014) Defining and observing stages of climate-mediated range shifts in marine systems. *Global Environmental Change* **26**, 27–38.

Behrenfeld MJ, O'Malley RT, Siegel DA, McClain CR, Sarmiento JL, Feldman GC, Milligan AJ, Falkowski PG, Letelier RM, Boss ES (2006) Climate-driven trends in contemporary ocean productivity. *Nature* **444**, 752–755.

Berkeley SA, Hixon MA, Larson RJ, Love MS (2004) Fisheries sustainability via protection of age structure and spatial distribution of fish populations. *Fisheries* **29**, 23–32.

Bindoff NL, Willebrand J, Artale V, Cazenave A, Gregory J, Gulev S, Hanawa K, Le Quéré C, Levitus S, Nojiri Y, Shum CK, Talley LD, Unnikrishnan A (2007) Observations: Oceanic Climate Change and Sea Level. In: Solomon, S., Qin D., Manning M., Chen Z., Marquis M., Averyt K. B., Tignor M., Miller H. L. (eds.), *Climate Change 2007: The Physical Science Basis. Contribution of Working Group I to the Fourth Assessment Report of the*

Intergovernmental Panel on Climate Change. Cambridge University Press, Cambridge, United Kingdom and New York, NY, USA.

Block BA, Teo SL, Walli A, Boustany A, Stokesbury MJ, Farwell CJ, Weng KC (2005) Electronic tagging and population structure of Atlantic bluefin tuna. *Nature* **434**, 1121–1127.

Boyce DG, Lewis MR, Worm B (2010) Global phytoplankton decline over the past century. *Nature* **466**, 591–596.

Buisson L, Grenouillet G, Villéger S, Canal J, Laffaille P (2013) Toward a loss of functional diversity in stream fish assemblages under climate change. *Global Change Biol.* **19**, 387–400.

Burt JA, Feary DA, Bauman AG, Usseglio P, Cavalcante GH, Sale PF (2011) Biogeographic patterns of reef fish community structure in the northeastern Arabian Peninsula. *ICES Journal of Marine Science* **68**, 1875–1883.

Byrne RH, Mecking S, Feely RA, Liu X (2010) Direct observations of basin-wide acidification of the North Pacific Ocean. *Geophysical Research Letters* **37**, L02601, doi:http://dx.doi.org/10.1029/2009GL040999.

Cai W, Borlace S, Lengaigne M, van Rensch P, Collins M, Vecchi G, Timmermann A, Santoso A, McPhaden MJ, Wu L, England MH, Wang G, Guilyardi E, Jin F-F (2014) Increasing frequency of extreme El Niño events due to greenhouse warming. *Nature Climate Change* **4**, 111–116.

Cairns DK, Gaston AJ, Huettmann F (2008) Endothermy, ectothermy and the global structure of marine vertebrate communities. *Marine Ecology Progress Series* **356**, 239–250.

Cavanaugh KC, Kellnerb JR, Fordec AJ, Grunerd DS, Parkera JD, Rodrigueza W, Fellera IC (2014) Poleward expansion of mangroves is a threshold response to decreased frequency of extreme cold events. *Proceedings of the National Academy of Sciences USA* **111**, 723–727.

Chavez FP, Ryan J, Lluch-Cota SE, Ñiquen M (2003) From anchovies to sardines and back: Multidecadal change in the Pacific Ocean. *Science* **299**, 217–221.

Checkley Jr. DM, Asch RG, Rykaczewski RR (2017) Climate, Anchovy, and Sardine. *Annual Review of Marine Science* **9**, 469–493.

Cheng L, Zhu J (2018) 2017 was the warmest year on record for the global ocean. *Advances in Atmospheric Sciences* **35**, 261–263.

Cheng L, Trenberth KE, Fasullo J, Boyer T, Abraham J, Zhu J (2017) Improved estimates of ocean heat content from 1960 to 2015. *Science Advances* **3**, e1601545. doi: 10.1126/sciadv.1601545.

Cheung WWL, Watson R, Pauly D (2012) Climate change induced tropicalization of marine communities in Western Australia. *Marine and Freshwater Research* **63**, 415–427.

Chust G, Allen JI, Bopp L, Schrum C, Holt J, Tsiaras K, Zavatarelli M, Chifflet M, Cannaby H, Dadou I, Daewel U, Wakelin SL, Machu E, Pushpadas D, Butenschon M, Artioli Y, Petihakis G, Smith C, Garçon V, Goubanova K, Le Vu B, Fach BA, Salihoglu B, Clementi E, Irigoien X (2014) Biomass changes and trophic amplification of plankton in a warmer ocean. *Global Change Biology* **20**, 2124–2139.

Church JA, Clark PU, Cazenave A, Gregory JM, Jevrejeva S, Levermann A, Merrifield MA, Milne GA, Nerem RS, Nunn PD, Payne AJ, Pfeffer WT, Stammer D, Unnikrishnan AS (2013) Sea level change. In: Stocker T. F., Qin D., Plattner G.-K. Tignor M., Allen S. K., Boschung J., Nauels A., Xia Y., Bex V., and Midgley P. M. (eds.), *Climate Change 2013: The Physical Science Basis. Contribution of Working Group I to the Fifth Assessment Report of the Intergovernmental Panel on Climate Change.* Cambridge University Press, Cambridge, United Kingdom and New York, USA.

Clark PU, Pisias NG, Stocker TF, Weaver AJ (2002) The role of the thermohaline circulation in abrupt climate change. *Nature* **415**, 863–869.

Collins M, Knutti R, Arblaster J, Dufresne J-L, Fichefet T, Friedlingstein P, Gao X, Gutowski WJ, Johns T, Krinner G, Shongwe M, Tebaldi C, Weaver AJ, Wehner M (2013) Long-term climate change: Projections, commitments and irreversibility, In: Collins M. *et al.*, (ed.), *Climate Change 2013: The Physical Science Basis. Contribution of Working Group I to the Fifth Assessment Report of the Intergovernmental Panel on Climate Change,* pp. 1029–1136, Cambridge University Press, Cambridge, UK and New York.

Conover DO, Munch SB (2002) Sustaining fisheries yields over evolutionary time scales. *Science* **297**, 95–96.

Cook J, Oreskes N, Doran PT, Anderegg WRL, Verheggen B, Maibach EW, Carlton JS, Lewandowsky S, Skuce AG, Green SA, Nuccitelli D, Jacobs P, Richardson M, Winkler B, Painting R, Rice K (2016) Consensus on consensus: A synthesis of consensus estimates on human-caused global warming. *Environmental Research Letters* **11**, doi:10.1088/1748-9326/11/4/048002.

Costanza R, de Groot R, Sutton P, van der Ploeg S, Anderson SJ, Kubiszewski I, Farber S, Turner RK (2014) Changes in the global value of ecosystem services. *Global Environmental Change* **26**, 152–158.

Cowtan K, Way RG (2014) Coverage bias in the HadCRUT4 temperature series and its impact on recent temperature trends. *Quarterly Journal of the Royal Meteorological Society* **140**, 1935–1944.

Czarnoleski M, Ejsmont-Karabin J, Angilletta MJ, Kozlowski J (2015) Colder rotifers grow larger but only in oxygenated waters. *Ecosphere* **6**, art164 10.1890/ES15-00024.1.

Daufresne M, Lengfellner K, Sommer U (2009) Global warming benefits the small in aquatic ecosystems. *Proceedings of the National Academy of Sciences of the United States of America* **106**, 12788–12793.

DeCarlo TM, Cohen AL, Wong GTF, Davis KA, Lohmann P, Soong K (2017) Mass coral mortality under local amplification of 2°C ocean warming. *Scientific Reports* **7**, Article No. 44586.

Dieng HB, Cazenave A, Meyssignac B, von Schuckmann K, Palanisamy H (2017) Sea and land surface temperatures, ocean heat content, Earth's energy imbalance and net radiative forcing over the recent years. *International Journal of Climatology* **37** (Suppl. 1), 218–229.

Donelson JM, Munday PL (2015) Transgenerational plasticity mitigates the impact of global warming to offspring sex ratios. *Global Change Biology* **21**, 2954–2962.

Donelson JM, Munday PL, McCormick MI, Pitcher CR (2012) Rapid transgenerational acclimation of a tropical reef fish to climate change. *Nature Climate Change* **2**, 30–32.

Donner SD, Skirving WJ, Little CM, Oppenheimer M, Hoegh-Guldberg O (2005) Global assessment of coral bleaching and required rates of adaptation under climate change. *Global Change Biology* **11**, 2251–2265.

Dorresteijn I, Kitaysky AS, Barger C, Benowitz-Fredericks ZM, Byrd GV, Shultz M, Young R (2012) Climate affects food availability to planktivorous least auklets *Aethia pusilla* through physical processes in the southeastern Bering Sea. *Marine Ecology Progress Series* **454**, 207–220.

Dulvy NK, Rogers SI, Jennings S, Stelzenmüller V, Dye SR, Skjoldal HR (2008) Climate change and deepening of the North Sea fish assemblage: A biotic indicator of warming seas. *Journal of Applied Ecology* **45**, 1029–1039.

Edwards M (2016) Impacts and effects of ocean warming on plankton. In: Laffoley D., Baxter J. M. (eds.), 2016. *Explaining Ocean Warming: Causes, Scale, Effects And Consequences*, Full Report, Gland, Switzerland: IUCN. pp. 75–86.

Eggert A, Burger EM, Breeman AM (2005) Ecotypic differentiation in thermal traits in the tropical to warm-temperate green macrophyte *Valonia utricularis*. *Botanica Marina* **46**, 69–81.

Eliason EJ, Clark TD, Hinch SG, Farrell AP (2011) Cardiorespiratory collapse at high temperature in swimming adult sockeye salmon. *Conservation Physiology* **1**, doi: 10.1093/conphys/cot008.

Eppley RW (1972) Temperature and phytoplankton growth in the sea. *Fishery Bulletin* **70**, 1063–1085.

Erwin CA, Congdon BC (2007) Day-to-day variation in sea-surface temperature negatively impacts sooty tern (*Sterna fuscata*) foraging success on the Great Barrier Reef, Australia. *Marine Ecology Progress Series* **331**, 255–266

Falkowski PG, Oliver MJ (2007) Mix and match: How climate selects phytoplankton. *Nature Reviews Microbioloyqy* **5**, 813–819.

Falkowski PG, Barber RT, Smetacek V (1998) Biogeochemical controls and feedbacks on ocean primary production. *Science* **281**, 200–206.

Feary DA, Burt JA, Bauman AG, Usseglio P, Sale PF, Cavalcante GH (2010) Fish communities on the world's warmest reefs: What can they tell us about the effects of climate change in the future? *Journal of Fish Biology* **77**, 1931–1947.

Findlay HS, Kendall MA, Spicer JI, Widdicombe S (2010) Post-larval development of two intertidal barnacles at elevated CO_2 and temperature. *Marine Biology* **157**, 725–735.

Forster J, Hirst AG, Atkinson D (2012) Warming-induced reductions in body size are greater in aquatic than terrestrial species. *Proceedings of the National Academy of Sciences of the United States of America* **109**, 19310–19314.

Fossheim M, Primicerio R, Johannesen E, Ingvaldsen RB, Aschan MM, Dolgov AV (2015) Recent warming leads to a rapid borealization of fish communities in the Arctic. *Nature Climate Change* **5**, 673–677.

Fry FEJ, Hart JS (1948) The relation of temperature to oxygen consumption in the gold-fish. *Biology Bulletin* **94**, 66–77.

Ganachaud A, Wunsch C (2000) Improved estimates of global oceanic circulation, heat transport and mixing from hydrographic data. *Nature* **408**, 453–457.

Gardner JL, Peters A, Kearney MR, Joseph L, Heinsohn R (2011) Declining body size: A third universal response to warming? *Trends in Ecology and Evolution* **26**, 285–291.

Gardiner NM, Munday PL, Nilsson GE (2010) Counter-gradient variation in respiratory performance of coral reef fishes at elevated temperatures. *PLoS ONE* **5**, e13299.

Gaston AJ (2001) Taxonomy and conservation: Thoughts on the latest BirdLife International listings for seabirds. *Marine Ornithology* **29**, 1–6.

Gilmour JP, Smith LD, Heyward AJ, Baird AH, Pratchett MS (2013) Recovery of an isolated coral reef system following severe disturbance. *Science* **340**, 69–71.

Gorman JB, Erdmann ES, Pickering BC, Horne PJ, Blum JR, Lucas HM, Patterson-Fraser DL, Fraser WR (2010) A new high-latitude record for the macaroni penguin (*Eudyptes chrysolophus*) at Avian Island, Antarctica. *Polar Biology* **33**, 1155–1158.

Hancke K, Glud RN (2004) Temperature effects on respiration and photosynthesis in three diatom dominated benthic communities. *Aquatic Microbial Ecology* **37**, 265–281.

Hansen J, Satoa M, Ruedy R, Schmidt GA, Lo K (2015) Global Temperature in 2014 and 2015. Available at http://www.wijstoppensteenkool.nl/wp-content/uploads/2015/01/20150116_Temperature2014james-hansen.pdf.

Hartmann, DL et al. (2013), Observations: Atmosphere and surface, In: Stocker T. F., Qin D., Plattner G.-K., Tignor M., Allen S. K., Boschung J., Nauels A., Xia Y., Bex V., Midgley P. M. (eds.), *Climate Change 2013: The Physical Science Basis. Contribution of Working Group I to the Fifth Assessment Report of the Intergovernmental Panel on Climate Change*, pp. 159–254, Cambridge University Press, Cambridge, UK.

Hegerl GC, Hanlon H, Beierkuhnlein C (2011) Climate science: Elusive extremes. *Nature Geoscience* **4**, 142–143.

Helm KP, Bindoff NL, Church JA (2011) Observed decreases in oxygen content of the global ocean. *Geophysical Research Letters* **38**, doi:10.1029/2011GL049513.

Henson SA, Sarmiento JL, Dunne JP, Bopp L, Lima I, Doney SC, John J, Beaulieu C (2010) Detection of anthropogenic climate change in satellite records of ocean chlorophyll and productivity. *Biogeosciences* **7**, 621–640.

Heron SF, Maynard JA, van Hooidonk R, Eakin CM (2016a) Warming trends and bleaching stress of the World's coral reefs 1985–2012. *Scientific Reports* **6**, 38402.

Heron SF, Eakin CM, Maynard JA, van Hooidonk R (2016b) Impacts and effects of ocean warming on coral reefs. In: Laffoley D., Baxter J. M. (eds.), *2016*.

Explaining Ocean Warming: Causes, Scale, Effects And Consequences. Full Report. Gland, Switzerland: IUCN. pp. 177–197.

Higini K, Sommer U (2013) Phytoplankton cell size reduction in response to warming mediated by nutrient limitation. *PLoS One* **8**(9), e71528.

Hoegh-Guldberg O, Jones RJ, Ward S, Loh WK (2002) Communication arising. Is coral bleaching really adaptive? *Nature* **415**, 601–602.

Hoegh-Guldberg O, Mumby PJ, Hooten AJ, Steneck RS, Greenfield P, Gomez E, Harvell CD, Sale PF, Edwards AJ, Caldeira K, Knowlton N, Eakin CM, Iglesias-Prieto R, Muthiga N, Bradbury RH, Dubi A, Hatziolos ME (2007) Coral reefs under rapid climate change and ocean acidification. *Science* **318**, 1737–1742.

Hoegh-Guldberg O, Cai R, Poloczanska ES, Brewer PG, Sundby S, Hilmi K, Fabry VJ, Jung S (2014) The ocean. In: Field C. B., Barros V. R., Dokken D. J., Mach K. J., Mastrandrea M. D., Bilir T. E., Chatterjee M., Ebi K. L., Estrada Y. O., Genova R. C., et al. (eds), *Climate Change 2014: Impacts, Adaptation and Vulnerability. Contribution of Working Group II to the Fifth Assessment Report of the Intergovernmental Panel on Climate Change.* Cambridge University Press, Cambridge, UK and New York, NY, USA. pp. 1655–1731.

Howells EJ, Beltran VH, Larsen NW, Bay LK, Willis BL, van Oppen MJH (2011) Coral thermal tolerance shaped by local adaptation of photosymbionts. *Nature Climate Change* **2**, 116–120.

Hughes TP, Kerry JT, Álvarez-Noriega M, Álvarez-Romero JG, Anderson KD, Baird AH, Babcock RC, Beger M, Bellwood DR, Berkelmans R, Bridge TC, Butler IR, Byrne M, Cantin NE, Comeau S, Connolly SR, Cumming GS, Dalton SJ, Diaz-Pulido G, Eakin CM, Figueira WF, Gilmour JP, Harrison HB, Heron SC, Hoey AS, J-PA Hobbs, Hoogenboom MO, Kennedy EV, Kuo C-Y, Lough JM, Lowe RJ, Liu G, McCulloch MT, Malcolm HA, McWilliam MJ, Pandolfi JM, Pears RJ, Pratchett MS, Schoepf V, Simpson T, Skirving WJ, Sommer B, Torda G, Wachenfeld DR, Willis BL, Wilson SK (2017) Global warming and recurrent mass bleaching of corals. *Nature* **543**, 373–377.

Hunt Jr GL, Stabeno P, Walters G, Sinclair E, Brodeure RD, Napp JM, Bond NA (2002) Climate change and control of the southeastern Bering Sea pelagic ecosystem. *Deep-Sea Research II* **49**, 5821–5853.

Ikeda T, Kanno Y, Ozaki K, Shinada A (2001) Metabolic rates of epipelagic marine copepods as a function of body mass and temperature. *Marine Biology* **139**, 587–596.

Intergovernmental Panel on Climate Change (IPCC) (2013) Climate Change 2013: The Physical Science Basis. Contribution of Working Group I to the Fifth Assessment Report of the Intergovernmental Panel on Climate Change, Cambridge University Press, Cambridge, UK, 1535 p.

Intergovernmental Panel on Climate Change (IPCC), "Climate Change 2014: Synthesis Report. Contribution of Working Group 32. Groups I, II and III to the Fifth Assessment Report of the Intergovernmental Panel on Climate Change" (IPCC, 2014); www.ipcc.ch/report/ar5/syr/.

Jentsch A, Kreyling J, Beierkuhnlein C (2007) A new generation of climate-change experiments: Events, not trends. *Frontiers in Ecology and the Environment* **5**, 365–374.

Johansen JL, Jones GP (2011) Increasing ocean temperature reduces the metabolic performance and swimming ability of coral reef damselfishes. *Global Change Biology* **17**, 2971–2979.

Johansen JL, Messmer V, Coker DJ, Hoey AS, Pratchett MS (2014) Increasing ocean temperatures reduce activity patterns of a large commercially important coral reef fish. *Global Change Biology* **20**, 1067–1074.

Kaplan A, Cane M, Kushnir Y, Clement A, Blumenthal M, Rajagopalan B (1998) Analyses of global sea surface temperature 1856–1991. *Journal of Geophysical Research* **103**, 567–589.

Kayanne H, Harii S, Ide Y, Akimoto F (2002) Recovery of coral populations after the 1998 bleaching on Shiraho Reef, in the southern Ryukyus, NW Pacific. *Marine Ecology Progress Series* **239**, 93–103.

Keeling RF, Körtzinger, A, Gruber N (2010) Ocean deoxygenation in a warming world. *Annual Review of Marine Science* **2**, 199–229.

Kitagawa T, Kimura S, Nakata H, Yamada H (2006) Thermal adaptation of Pacific bluefin tuna *Thunnus orientalis* to temperate waters. *Fisheries Science* **72**, 149–156.

Kordas, RL, CDG Harley, and MI O'Connor 2011. Community ecology in a warming world: The influence of temperature on interspecific interactions in marine systems. *Journal of Experimental Marine Biology and Ecology* **400**, 218–226.

Kroeker KJ, Kordas RL, Crim R, Hendriks I, Ramajos L, Singh G, Duarte CM, Gattuso J-P (2013) Impacts of ocean acidification on marine organisms: Quantifying sensitivities and interaction with warming. *Global Change Biology* **19**, 1884–1896.

Laufkötter C, Vogt M, Gruber N, Aita-Noguchi M, Aumont O, Bopp L, Buitenhuis E, Doney SC, Dunne J, Hashioka T, Hauck J, Hirata T, John J, Le Quéré C, Lima ID, Nakano H, Seferian R, Totterdell I, Vichi M, Völker C (2015) Drivers and uncertainties of future global marine primary production in marine ecosystem models. *Biogeosciences* **12**, 6955–6984.

Le Quéré C, Moriarty R, Andrew RM, Canadell JG, Sitch S, Korsbakken JI, Friedlingstein P, Peters GP, Andres RJ, Boden TA, Houghton RA, House JI, Keeling RG, Tans P, Arneth A, Bakker DCE, Barbero L, Bopp L, Chang J, Chevallier F, Chini LP, Ciais P, Fader M, Feely RA, Gkritzalis T, Harris I, Hauck J, Ilyina T, Jain AK, Kato E, Kitidis V, Klein Goldewijk K, Koven C, Landschützer P, Lauvset SK, Lefèvre N, Lenton A, Lima ID, Metzl N, Millero F, Munro DR, Murata A, Nabel JEMS, Nakaoka S, Nojiri Y, O'Brien K, Olsen A, Ono T, Pérez FF, Pfeil B, Pierrot D, Poulter B, Rehder G, Rödenbeck C, Saito S, Schuster U, Schwinger J, Séférian R, Steinhoff T, Stocker BD, Sutton AJ, Takahashi J, Tilbrook B, van der Laan-Luijkx IT, van der Werf GR, van Heuven S, Vandemark D, Viovy N, Wiltshire A, Zaehle S, Zeng N (2015) Global carbon budget 2015. *Earth System Science Data* **7**, 349–396.

Levitus S, Antonov JI, Boyer TP, Stephens C (2000) Warming of the world ocean. *Science* **287**, 2225–2229.

Levitus S, Antonov J, Boyer T (2005). Warming of the world ocean, 1955–2003, *Geophysical Research Letters* **32**, L02604, doi:10.1029/2004GL021592.

Lewandowska AM, Boyce DG, Hofmann M, Matthiessen B, Sommer U, Worm B (2014) Effects of sea surface warming on marine plankton. *Ecology Letters* **17**, 614–623.

Liang X, Wunsch C, Heimbach P, Forget G (2015) Vertical redistribution of oceanic heat content. *J. Climate* **28**, 3821–3833.

Loeb V, Siegel V, Holm-Hansen O, Hewitt R, Fraserk W, Trivelpiecek W, Trivelpiecek S (1997) Effects of sea-ice extent and krill or salp dominance on the Antarctic food web. *Nature* **387**, 897–900.

López-Urrutia A, San Martin E, Harris RP, Irigoien X (2006) Scaling the metabolic balance of the oceans. *Proceedings of the National Academy of Sciences USA* **103**, 8739–8744.

Manzello DP, Eakin CM, Glynn PW (2017) Effects of global warming and ocean acidification on carbonate budgets of Eastern Pacific coral reefs. In: Glynn P. W., Manzello D. P., Enochs I. C. (eds.), *Coral Reefs of the Eastern Tropical Pacific: Persistence and Loss in a Dynamic Environment — Coral Reefs of the World* **8**, 517–533.

Masson-Delmotte V, Schulz M, Abe-Ouchi A et al. (2013) Information from paleoclimate archives. In: Stocker T. F., Qin D., Plattner G.-K., et al. (eds.), *Climate Change 2013: The Physical Science Basis. Contribution of Working Group I to the Fifth Assessment Report of the Intergovernmental Panel on Climate.* Change Cambridge University Press, Cambridge. pp. 383–464.

Mather FJ, Mason JM, Jones AC (1995) Historical document: Life history and fisheries of Atlantic bluefin tuna. National Oceanic and Atmospheric Administration Technical Memorandum 370, National Marine Fisheries Service, Southeast Fisheries Science Center, 174 p.

McIntyre T, Ansorge IJ, Bornemann H, Plötz J, Tosh CA, Bester MN (2011) Elephant seal dive behaviour is influenced by ocean temperature: Implications for climate change impacts on an ocean predator. *Marine Ecology Progress Series* **441**, 257–272.

Millien V, Lyons SK, Olson L, Smith FA, Wilson AB, Yom-Tov Y (2006) Ecotypic variation in the context of global climate change: Revisiting the rules. *Ecology Letters* **9**, 853–869.

Mojica KDA, Huisman J, Wilhelm SW, Brussaard CPD (2015) Latitudinal variation in virus-induced mortality of phytoplankton across the North Atlantic Ocean. *ISME Journal* **10**, 500–513.

Montero-Serra I, Edwards M, Genner MJ (2015) Warming shelf seas drive the subtropicalization of European pelagic fish communities. *Global Change Biology* **21**, 144–153.

Moran XAG, Lopez-Urrutia A, Calvo-Diaz A, Li WKW (2010) Increasing importance of small phytoplankton in a warmer ocean. *Global Change Biology* **16**, 1137–1144.

Muhling BA, Brill R, Lamkin JT, Roffer MA, Lee S-K, Liu Y, Muller-Karger F (2017) Projections of future habitat use by Atlantic bluefin tuna: Mechanistic vs. correlative distribution models. *ICES Journal of Marine Science* **74**, 698–716.

Munday PL, Kingsford MJ, O'Callaghan M, Donelson JM (2008) Elevated temperature restricts growth potential of the coral reef fish *Acanthochromis polyacanthus*. *Coral Reefs* **27**, 927–931.

Munday PL, McCormick MI, Nilsson GE (2012) Impact of global warming and rising CO_2 levels on coral reef fishes: What hope for the future? *Journal of Experimental Biology* **215**, 3865–3873.

Munday PL, Donelson JM, Domingos JA (2017) Potential for adaptation to climate change in a coral reef fish. *Global Change Biology* **23**, 307–317.

Munoz NJ, Farrell AP, Heath JW, Neff BD (2015) Adaptive potential of a Pacific salmon challenged by climate change. *Nature Climate Change* **5**, 163–166.

Nicholls RJ, Cazenave A (2010) Sea-level rise and its impact on coastal zones. *Science* **328**, 1517–1520.

Nilsson GE, Crawley N, Lunde IG, Munday PL (2009) Elevated temperature reduces the respiratory scope of coral reef fishes. *Global Change Biology* **15**, 1405–1412.

NOAA National Centers for Environmental Information (2016) State of the Climate: Global Analysis for Annual 2015. Published Online January 2016. Retrieved on January 28, 2016 from http://www.ncdc.noaa.gov/sotc/global/201513.

Ochoa-Ochoa LM, Rodriguez P, Mora F, Flores-Villela O, Whittaker RJ (2012) Climate change and amphibian diversity patterns in Mexico. *Biological Conservation* **150**, 94–102.

Ohlberger J (2013) Climate warming and ectotherm body size — from individual physiology to community ecology. *Functional Ecology* **27**, 991–1001.

O'Leary JK, Micheli F, Airoldi L, Boch C, de Leo G, Elahi R, Ferretti F, Graham NAJ, Litvin SY, Low NH, Lummis S, Nickols KJ, Wong J (2017) The resilience of marine ecosystems to climatic disturbances. *BioScience* **67**, 208–220.

Oliver TH, Heard MS, Isaac NJB, Roy DB, Procter D, Eigenbrod F, Freckleton R, Hector A, Orme CDL, Petchey OL, Proença V, Raffaelli D, Suttle KB, Mace GM, Martín-López B, Woodcock BA, Bullock JM (2015) Biodiversity and resilience of ecosystem functions. *Trends in Ecology and Evolution* **30**, 673–684.

Pagano AM, Durner GM, Rode KD, Atwood TC, Atkinson N, Peacock E, Costa DP, Owen MA, Williams TM (2018) High-energy, high-fat lifestyle challenges an Arctic apex predator, the polar bear. *Science* **359**, 568–572.

Pankhurst NW, Munday PL (2011) Effects of climate change on fish reproduction and early life history stages. *Marine and Freshwater Research* **62**, 1015–1026.

Passow U, Carlson CA (2012) The biological pump in a high CO_2 world. *Marine Ecology Progress Series* **470**, 249–271.

Paul MJ, Zucker I, Schwartz WJ (2008) Tracking the seasons: The internal calendars of vertebrates. *Philosophical Transactions of the Royal Society B* **363**, 341–361.

Pecl GT, *et al.* (2017) Biodiversity redistribution under climate change: Impacts on ecosystems and human well-being. *Science* **355**, doi: 10.1126/science.aai9214.

Peron C, Weimerskirch H, Bost C-A (2012) Projected poleward shift of king penguins' (*Aptenodytes patagonicus*) foraging range at the Crozet Islands, southern Indian Ocean. *Proceedings of the Royal Society B Biological Sciences* **279**, 2515–2523.

Perry AL, Low PJ, Ellis JR, Reynolds JD (2005) Climate change and distribution shifts in marine fishes. *Science* **308**, 1912–1915.

Petry MV, Basler AB, Leal Valls FC, Krüger L (2013) New southerly breeding location of king penguins (*Aptenodytes patagonicus*) on Elephant Island (Maritime Antarctic). *Polar Biology* **36**, 603–606.

Pinsky ML, Worm B, Fogarty MJ, Sarmiento JL, Levin SA (2013) Marine taxa track local climate velocities. *Science* **341**, 1239–42

Porch CE, Lauretta MV (2016) On making statistical inferences regarding the relationship between spawners and recruits and the irresolute case of western Atlantic bluefin tuna (*Thunnus thynnus*). *PLoS One* **11**, e0156767.

Pörtner HO (2001) Climate change and temperature-dependent biogeography: Oxygen limitation of thermal tolerance in animals. *Naturwissenschaften* **88**, 137–146.

Pörtner HO, Knust R (2007) Climate change affects marine fishes through the oxygen limitation of thermal tolerance. *Science* **315**, 95–97.

Pörtner HO, Farrell AP (2008) Physiology and climate change. *Science* **322**, 690–692.

Prada F, Caroselli E, Mengoli S, Brizi L, Fantazzini P, Capaccioni B, Pasquini L, Fabricius KE, Dubinsky Z, Falini G, Goffredo S (2017) Ocean warming and acidification synergistically increase coral mortality, *Nature Scientific Reports* **7**, 40842 doi: 10.1038/srep40842.

Punzon A, Serrano A, Sanchez F, Velasco F, Preciado I, Gonzalez-Irusta JM, Lopez-Lopez L (2016) Response of a temperate demersal fish community to global warming. *Journal of Marine Systems* **161**, 1e10.

Purkey SG, Johnson GC (2010) Warming of global abyssal and deep Southern Ocean waters between the 1990s and 2000s: Contributions to global heat and sea level rise budgets. *Journal of Climate* **23**, 6336–6351.

Reid PC (2016) Ocean warming: Setting the scene. In: Laffoley D., Baxter J. M., (eds.), *Explaining Ocean Warming: Causes, Scale, Effects and Consequences.* Full Report, Gland, Switzerland: IUCN, 17–45.

Reid PC, Fischer AC, Lewis-Brown E, Meredith MP, Sparrow M, Andersson AJ, Anita A, Bates NR, Bathmann U, Beaugrand G, *et al.* (2009) Chapter 1, Impacts of the oceans on climate change. *Advances in Marine Biology* **56**, 1–150.

Reusch TBH (2014) Climate change in the oceans: evolutionary versus phenotypically plastic responses. *Evolutionary Applications* **7**, 104–122.

Reusch TBH, Ehlers A, Hämmerli A, Worm B (2005) Ecosystem recovery after climatic extremes enhanced by genotypic diversity. *Proceedings of the National Academy of Sciences* **102**, 2826–2831.

Rhein M, Rintoul SR, Aoki S, Campos E, Chambers D, Feely RA, et al. (2013) Observations: Ocean, in Climate Change 2013: The Physical Science Basis. In: Stocker T. F., Qin D., Plattner G.-K., Tignor M., Allen S. K., Boschung J., Nauels A., Xia Y., Bex V., Midgley P. M. Contribution of Working Group I to the Fifth Assessment Report of the Intergovernmental Panel on Climate Change (eds.), (Cambridge; New York, NY: Cambridge University Press), 255–316.

Richardson TL, Jackson GA (2007) Small phytoplankton and carbon export from the surface ocean. *Science* **315**, 838–840.

Roberts CD, Palmer MD, Mcneall D, Collins M (2015) Quantifying the likelihood of a continued hiatus in global warming. *Nature Climate Change* **5**, 337–342.

Rodolfo-Metalpa R, Houlbreque F, Tambutte E, Boisson F, Baggini C, Patti FP, Jeffree R, Fine M, Foggo A, Gattuso J-P, Hall-Spencer JM (2011) Coral and mollusc resistance to ocean acidification adversely affected by warming. *Nature Climate Change* **1**, 308–312.

Roemmich D, Gould WJ, Gilson J (2012) 135 years of global ocean warming between the Challenger expedition and the Argo Programme. *Nature Climate Change* **2**, 425–428.

Ruhl HA, Smith Jr. KL (2004) Shifts in deep-sea community structure linked to climate and food supply. *Science* **305**, 513–515.

Rummer JL, Munday PL (2017) Climate change and the evolution of reef fishes: Past and future. *Fish and Fisheries* **18**, 22–39.

Rummer JL, Couturier CS, Stecyk JAW et al. (2014) Life on the edge: Thermal optima for aerobic scope of equatorial reef fishes are close to current day temperatures. *Global Change Biology* **20**, 1055–1066.

Ruppel CD, Kessler JD (2017) The interaction of climate change and methane hydrates. *Reviews of Geophysics* **55**, 126–168.

Sabine CL, Feely RA, Gruber N, Key RM, Lee K, Bullister JL, Wanninkhof R, Wong CS, Wallace DWR, Tilbrook B, Millero FJ, Peng T-H, Kozyr A, Ono T, Rios AF (2004) The oceanic sink for anthropogenic CO_2. *Science* **305**, 367–371 (2004).

Salinas S, Munch SB (2012) Thermal legacies: Transgenerational effects of temperature on growth in a vertebrate. *Ecology Letters* **15**, 159–163.

Sand-Jensen K, Pedersen NL, Søndergaard M (2007) Bacterial metabolism in small temperate streams under contemporary and future climates. *Freshwater Biology* **52**, 2340–2353.

Schmittner A (2005) Decline of the marine ecosystem caused by a reduction in the Atlantic overturning circulation. *Nature* **434**, 628–633.

Smith KL, Ruhl HA, Kahru M, Huffard CL, Sherman AD (2013) Deep ocean communities impacted by changing climate over 24 years in the abyssal northeast Pacific Ocean. *Proceedings of the National Academy of Sciences* **110**, 19838–19841.

Smith SJ, Edmonds J, Hartin CA, Mundra A, Calvin K (2015) Near-term acceleration in the rate of temperature change. *Nature Climate Change* **5**, 333–336.

Sommer U, Lengfellner A (2008) Climate change and the timing, magnitude, and composition of the phytoplankton spring bloom. *Global Change Biology* **14**, 1199–1208.

Sommer U, Aberle N, Lengfellner K, Lewandowska A (2012) The Baltic Sea spring phytoplankton bloom in a changing climate: An experimental approach. *Marine Biology* **159**, 2479–2490.

Somero, G.N. (2010) The physiology of climate change: How potentials for acclimatization and genetic adaptation will determine 'winners' and 'losers'. *Journal of Experimental Biology* **213**, 912–920.

Sorte CJB, Fuller A, Bracken MES (2010) Impacts of a simulated heat wave on composition of a marine community. *Oikos* **119**, 1909–1918.

Stern NH (2007) *The Economics of Climate Change: The Stern Review* + Cambridge University Press, Cambridge, UK.

Stevens ED, Carey FG (1981) One why of the warmth of warm-bodied fish. *American Journal of Physiology — Regulatory, Integrative and Comparative Physiology* **240**, R151–R155.

Stillman JH (2003) Acclimation capacity underlies susceptibility to climate change. *Science* **301**, 65.

Stirling I, Derocher AE (2012) Effects of climate warming on polar bears: A review of the evidence. *Global Change Biology* **18**, 2694–2706.

Stramma L, Prince ED, Schmidtko S, Luo J, Hoolihan JP, Visbeck M, Wallace DWR, Brandt P, Körtzinger A (2012) Expansion of oxygen minimum zones may reduce available habitat for tropical pelagic fishes. *Nature Climate Change* **2**, 33–37.

Sunday JM, Bates AE, Dulvy NK (2011) Global analysis of thermal tolerance and latitude in ectotherms. *Proceedings of the Royal Society B* **278**, 1823–1830.

Sweetman AK, Thurber AR, Smith CR, Levin LA, Mora C, Wei C-L, Gooday AJ, Jones DOB, Rex M, Yasuhara M, Ingels J, Ruhl HA, Frieder CA, Danovaro R, Würzberg L, Baco A, Grupe BM, Pasulka A, Meyer KS, Dunlop KM, Henry L-A, Roberts JM (2017) Major impacts of climate change on deep-sea benthic ecosystems. *Elementa: Science of the Anthropocene* **5,** 4, doi: https://doi.org/10.1525/elementa.203.

Swingedouw D, Braconnot P, Marti O (2006) Sensitivity of the Atlantic Meridional overturning circulation to the melting from northern glaciers in climate change experiments. *Geophysical Research Letters* **33**, l07711, doi:10.1029/2006GL025765.

Tewksbury JJ, Huey RB, Deutsch CA (2008) Ecology — putting the heat on tropical animals. *Science* **320**, 1296–1297.

Trenberth KE, Dai A (2007) Effects of Mount Pinatubo volcanic eruption on the hydrological cycle as an analog of geoengineering. *Geophysical Research Letters* **34**, L15702, doi:10.1029/2007GL030524.

Trenberth KE, Stepaniak DP, Caron JM (2002) Interannual variations in the atmospheric heat budget. *Journal of Geophysical Research* **107**, D8, 4066, doi:10.1029/2000JD000297.

Trenberth KE, Fasullo JT, Balmaseda MA (2014) Earth's energy imbalance. *Journal of Climate* **27**, 3129–3144.

Verges A, Doropoulos C, Malcolm HA, Skye M, Garcia-Pizáa M, Marzinellia EM, Campbell AH, Ballesteros E, HoeyI AS, Vila-Concejo Aj, Bozec YM, Steinberg PD (2016) Long-term empirical evidence of ocean warming leading to tropicalization of fish communities, increased herbivory, and loss of kelp. *Proceedings of the National Academy of Sciences of the USA* **113**, 13791–13796.

Vindenes Y, Edeline E, Ohlberger J, Langangen O, Winfeld IJ, Stenseth NC, Vollestad LA (2014) Effects of climate change on trait-based dynamics of a top predator in freshwater ecosystems. *American Naturalist* **183**, 243–256.

von Schuckmann K, Palmer MD, Trenberth KE, Cazenave A, Chambers D, Champollion N, Hansen J, Josey SA, Loeb N, Mathieu P-P, Meyssignac B, Wild M (2016) An imperative to monitor Earth's energy imbalance. *Nature Climate Change* **6**, 138–144.

Ward TD, Algera DA, Gallagher AJ, Hawkins E, Horodysky A, Jørgensen C, Killen SS, McKenzie DJ, Metcalfe JD, Peck MA, Vu M, Cooke SJ (2016) Understanding the individual to implement the ecosystem approach to fisheries management. *Conservation Physiology* **4**, 1–10.

Ware JV, Rode KD, Bromaghin JF, Douglas DC, Wilson RR, Regehr EV, Amstrup SC, Durner GM, Pagano AM, Olson J, Robbins CT, Jansen HT (2017) Habitat degradation affects the summer activity of polar bears. *Oecologia* **184**, 87–99.

Wernberg T, Smale DA, Tuya F, Thomsen MS, Langlois TJ, de Bettignies T, Bennett S, Rousseaux CS (2013) An extreme climatic event alters marine ecosystem structure in a global biodiversity hotspot. *Nature Climate Change* **3**, 78–82.

Whitmarsh F, Zika J, Cazaja A (2015) Ocean heat uptake and the global surface temperature record, Grantham Institute, Briefing Paper No 14.

Wijffels S, Roemmich D, Monselesan D, Church J, Gilson J (2016) Ocean temperatures chronicle the ongoing warming of Earth. *Nature Climate Change* **6**, 116–118.

Wild M, Folini D, Schär C, Loeb N, Dutton EG, König-Langlo G (2013) The global energy balance from a surface perspective. *Climate Dynamics* **40**, 3107–3134.

Williams SE, Shoo LP, Isaac JL, Hoffmann AA, Langham G (2008) Towards an integrated framework for assessing the vulnerability of species to climate change. *PLOS Biol* **6**, 35, 2621–2626.

Wood CM, McDonald DG (1997) *Global Warming: Implications for Freshwater and Marine Fish*. Cambridge University Press, Cambridge.

Woodworth-Jefcoats PA, Polovina JJ, Drazen JC (2017) Climate change is projected to reduce carrying capacity and redistribute species richness in North Pacific pelagic marine ecosystems. *Global Change Biology* **23**, 1000–1008.

Yan X-H, Boyer T, Trenberth K, Karl TR, Xie S-P, Nieves V, Tung K-K, Roemmich D (2016) The global warming hiatus: Slowdown or redistribution? *Earth's Future* **4**, doi:10.1002/2016EF000417.

Chapter 4

Ocean Deoxygenation

Warming and Coastal Pollution Are Causing the Oceans to Become Oxygen-Starved

> "*Loss of oxygen in many ways is the destruction of an ecosystem. If we were creating vast areas on land that were uninhabitable by most animals, we'd notice. But we don't always see things like this when they are happening in the water.... There's potential for a feedback, where warming increases low-oxygen areas which produce nitrous oxide, which then causes more warming. That's a real concern.*"
>
> — Dr. Denise Breitburg, Senior Scientist at The Smithsonian Environmental Research Center, speaking on marine hypoxia.

Introduction

The transition from an anoxic atmosphere and hydrosphere to an oxic one appears to have occurred around 2 billion years ago. Today, oxygen constitutes around one-fifth of the atmosphere volume and it is essential for the survival of most marine animals and plants. Human activities, largely the burning of fossil fuels, are causing a decline in oxygen levels in the atmosphere although these changes are very small: 19 O_2 molecules are lost every year out of 1 million O_2 molecules in the atmosphere (Keeling, 2016). On the other hand, largely caused by warming and reduced ventilation of the ocean (Shepherd et al., 2017), oxygen gas (O_2) solubility is decreasing, causing oceanic deoxygenation and, in some cases, anoxia (lack of oxygen). Also, increased stratification of the surface ocean reduces O_2 supply to depth and thus the concentration of dissolved O_2 in the ocean interior (Sarmiento et al., 1998; Keeling and Garcia, 2002; Schmidtko et al., 2017). Indeed over the past five decades, oxygen levels in the ocean have declined by 2% (Schmidtko et al., 2017). This decrease in oxygen concentrations has been observed in coastal regions as well as in the open ocean, leading to increasing "dead zones", where hypoxia (low-oxygen conditions) triggers mortality events resulting, in some cases, in a depletion of fisheries (Rabalais et al., 2002). Also, as a result of increased stratification causing poor ventilation (slowing down of ocean circulation) of the ocean interior, there has been an expansion of the mid-depth zones with extremely low oxygen levels; these are termed oxygen-minimum zones (OMZs) (Cline and Richards, 1972; Keeling et al., 2010; Stramma et al., 2008). This chapter brings together a synthesis of the main impacts of ocean deoxygenation on marine life and the biogeochemical consequences of hypoxia in marine ecosystems.

Where Does Deoxygenation Occur in the Oceans?

In coastal waters and estuaries, hypoxic (O_2 concentrations, typically <60 μmol/kg, detrimental to most marine organisms), suboxic (extremely low-O_2 conditions which are life-threatening to most marine life, typically <5 μmol/kg), and anoxic (lack of oxygen) conditions are frequent. Anoxia also occurs in isolated enclosed basins, such as the Baltic Sea and the Black Sea. In coastal environments, the main causes of coastal deoxygenation are high photosynthetic rates and biomass production that fuel high rates of O_2 consumption by bacteria in subsurface waters and sediments. Coastal waters also lose O_2 from eutrophication by agricultural runoff or sewage discharge, leading to the excessive production of organic matter that increases oxygen demand in coastal ecosystems (Service, 2004). Specifically, eutrophication causes a drop in dissolved oxygen levels in bottom waters, driven by the enhanced growth of phytoplankton and subsequent death, downward flow to the ocean floor and, ultimately, organic matter decomposition by bacteria. As a result, the formation of dead zones is particularly widespread in coastal areas fueled by riverine runoff of fertilizers. Oxygen levels in coastal waters can also be affected by the larger-scale oxygen changes in the global ocean as O_2 supply to coastal waters also occurs via seawater exchange with the open ocean (Schmidtko et al., 2017).

Major concerns about ocean deoxygenation are its detrimental effects to marine life because the majority of marine animals and plants (photosynthetic organisms use oxygen in the dark) require oxygen. Hypoxia also leads to major declines in biodiversity, and many organisms experience sublethal stress, causing reduced growth and reproduction, physiological disruption, forced migration, and increased susceptibility to predation among many other impacts (Rabalais et al., 2002). Perhaps, the most susceptible organisms are those living on/in the benthos[1] because coastal sediments tend to become depleted in oxygen as they live far from the surface ocean, which is in contact with atmospheric oxygen. Additionally, decreasing dissolved oxygen concentrations in seawater affect marine biogeochemical processes, particularly those governing the global carbon and nitrogen cycles (Bange et al., 2005).

Hypoxia

Dissolved oxygen observations in seawater are available since 1903, and the first basin-wide surveys were conducted in the 1920s (Schmidtko et al., 2017). The global oceanic dissolved oxygen amounts to 227.4 ± 1.1 Pmol and it is non-uniformly distributed (Figures 1 and 2) (Schmidtko et al., 2017). High levels of O_2 can be found in the ventilated mixed-layer depth (below the surface) and also at greater depths (2,000–4,000 m) as a result of deep-water ventilation (Schmidtko et al., 2017).

However, multiple studies provide unequivocal evidence indicating an increase in the number of hypoxic zones and in their extension, severity, and duration (Stramma et al., 2008; Schmidtko et al., 2017). It is expected that this pattern will be maintained because, in addition to decreasing oxygen solubility in seawater, the continued warming of the oceans increases oxygen demand by organisms and limits ventilation of coastal waters, where dissolved oxygen has changed dramatically over the past decades, arguably more so than any other ecological stressor (Diaz, 2001; Vaquer-Sunyer and Duarte, 2008). Additionally, eutrophication of coastal systems, for example, by riverine runoff of fertilizers (Diaz and Rosenberg, 2008), leads to excessive production of organic matter, promoting removal of oxygen

---

[1] The biological community living on/in the seabed.

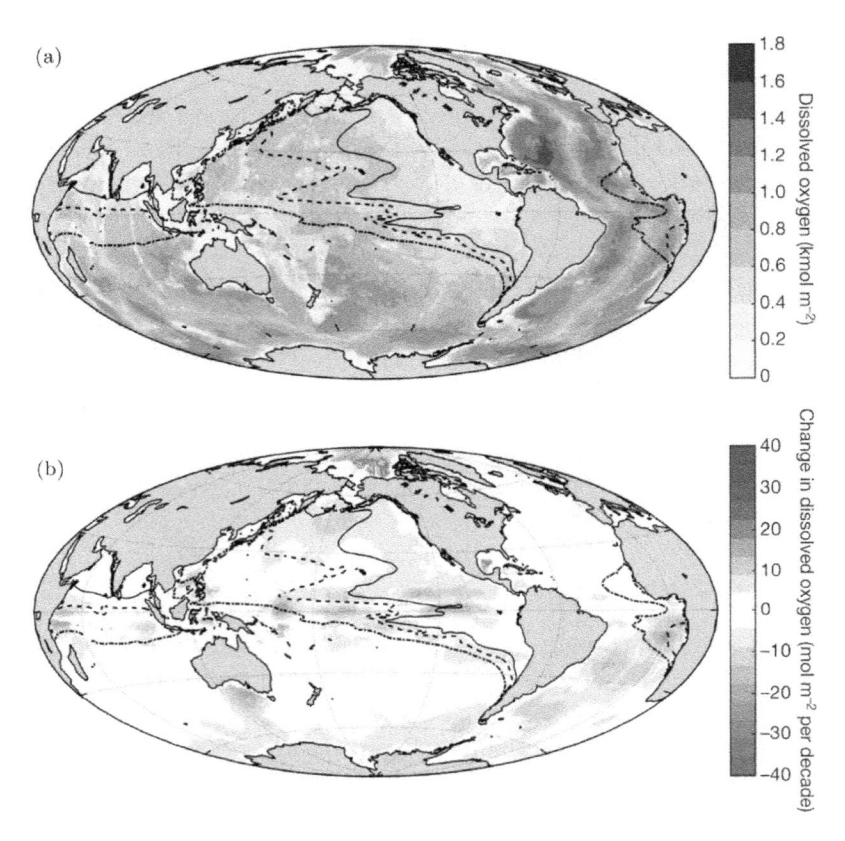

Figure 1. Dissolved oxygen and decadal changes since 1960: (a) global oxygen inventory (color bar represents dissolved oxygen) and (b) change in dissolved oxygen per decade (color coded). Lines indicate boundaries of OMZs: dashed-dotted lines represent regions with less than 80 μmol kg^{-1} oxygen anywhere within the water column; dashed lines and solid lines similarly represent regions with less than 40 μmol kg^{-1} oxygen and 20 μmol kg^{-1} oxygen, respectively (from Schmidtko et al., 2017).

by bacterial respiration that increases oxygen demand of the ecosystem (Vaquer-Sunyer and Duarte, 2008).

In the open ocean, deoxygenation is linked to warming-driven declines in oxygen solubility and reduced ventilation of deeper waters caused by enhanced upper-ocean stratification (Keeling et al., 2010). The regional long-term trends of oxygen indicate interannual and multi-decadal variability consistent with natural climate variability, for example, El Niño Southern Oscillation and the Pacific Decadal Oscillation (Stendardo and Gruber, 2012; Schmidtko et al., 2017). In the tropical and subtropical

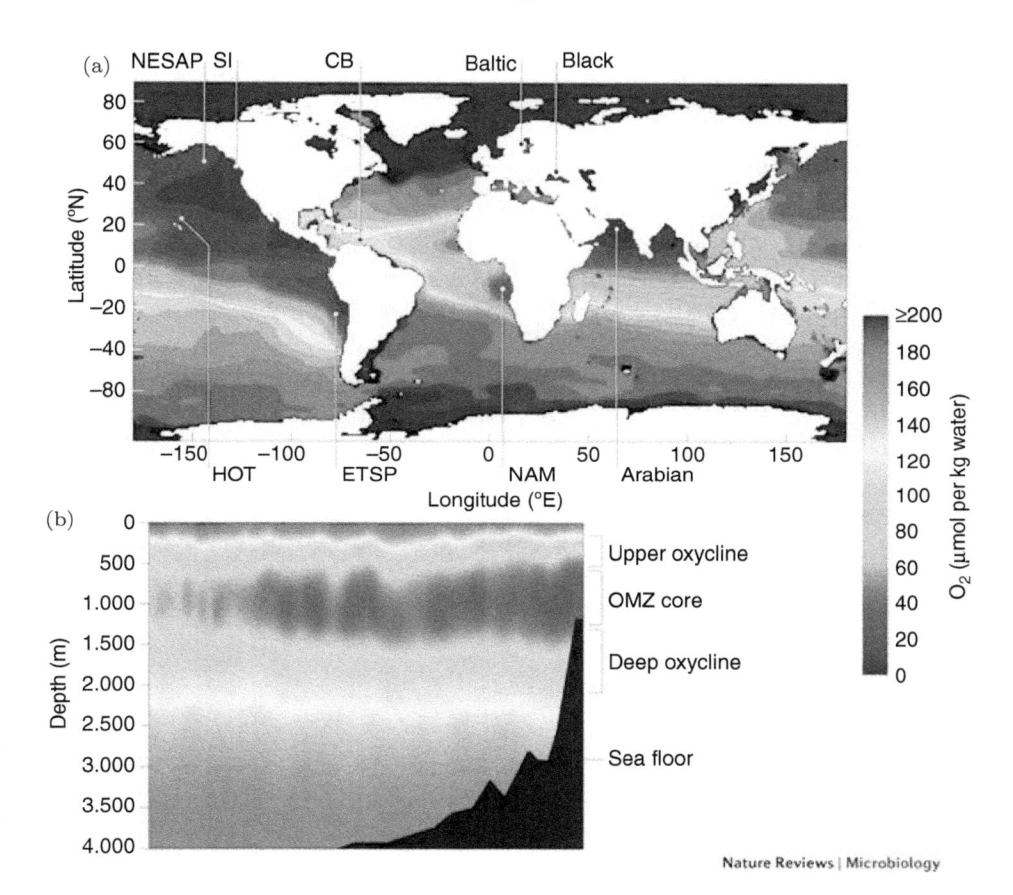

Figure 2. O_2 concentrations in the ocean: (a) Minimum O_2 concentrations for different regions of the ocean. Locations highlighted are the Hawaii Ocean Time-series (HOT), the northeast subarctic Pacific (NESAP), Saanich Inlet (SI), the eastern tropical South Pacific (ETSP), the Cariaco Basin (CB), the Namibian upwelling (NAM, also known as the Benguela upwelling), and the Baltic, Black and Arabian seas. (b) Cross-section of the NESAP OMZ, showing the O_2 concentration from surface waters to the sea floor. Upper oxycline: transition from surface waters to the OMZ core. OMZ core: defined by O_2 concentrations <20 μmol per kg water. Deep oxycline: transition from the bottom of the OMZ core to abyssal waters (from Wright et al., 2012).

latitudes, declines in O_2 in the upper ocean are attributed to the shoaling of the thermocline[2] depth, whereas for other latitudes, warming-driven

---

[2] A thermocline is the transition layer between the upper ocean mixed water and the cooler deep water below.

changes in solubility, wind forcing and large-scale ocean circulation appear to govern hypoxia (Deutsch et al., 2005; 2011).

Oxygen-Minimum Zones

Oxygen-Minimum Zones (OMZs) are formed because of weak ocean ventilation, which limits the supply of oxygen to depth, and elevated respiration, which consumes oxygen. The oxygen levels in OMZs are <20 μM although typically <1 μM (Karstensen et al., 2008), and waters of OMZs carry ~50 times less oxygen than the traditionally termed "O_2 minimum", typically found at intermediate depths (1,000–1,500 m) (Wyrtki, 1962). If we use the criterion that OMZs are <20 μM, they currently constitute 1–7% of the volume of the global ocean, occupying approximately 102 m km^3 (Wright et al., 2012). OMZs are also characterized by very low-pH (because CO_2 is released in remineralization[3]) and reduced conditions. OMZs are typically below the subsurface ocean, starting from 50 to 100 m depth in the Eastern South Pacific, Eastern Tropical North Pacific, Arabian Sea, and Bay of Bengal (Paulmier et al., 2008).

Biogeochemical Impacts of OMZs

Oxygen availability affects the remineralization of organic matter and, consequently, the cycling of nutrients through the water column. Within OMZs, low oxygen availability can affect the concentration of nutrients, including nitrogen, phosphorus, and iron. Therefore, decreases in oxygen availability can affect nutrient budgets and, consequently, productivity and carbon fixation on regional and global scales (Breitburg et al., 2018).

One example of the extent to which oxygen governs elemental cycles is the role of OMZs in the global nitrogen cycle. In the surface ocean, which is oxygenated, fixed nitrogen (N_2) largely ends up as nitrate ions (NO_3^-) (Figure 3). Also, under oxic conditions, bacteria decompose organic matter to produce ammonium (NH_4^+) and subsequently, nitrifying organisms transform NH_4^+ into nitrate (NO_3^-) while producing the greenhouse gas nitrous oxide (N_2O) as a byproduct (Santoro et al., 2011). On the other hand, in anoxic parts of the water column and in sediments, where dissolved O_2 is almost completely depleted, denitrification, the process by which NO_3^- (one of the main limiting nutrients in the ocean) is converted into N_2 by

---

[3] Remineralization is the breakdown of organic matter into inorganic compounds.

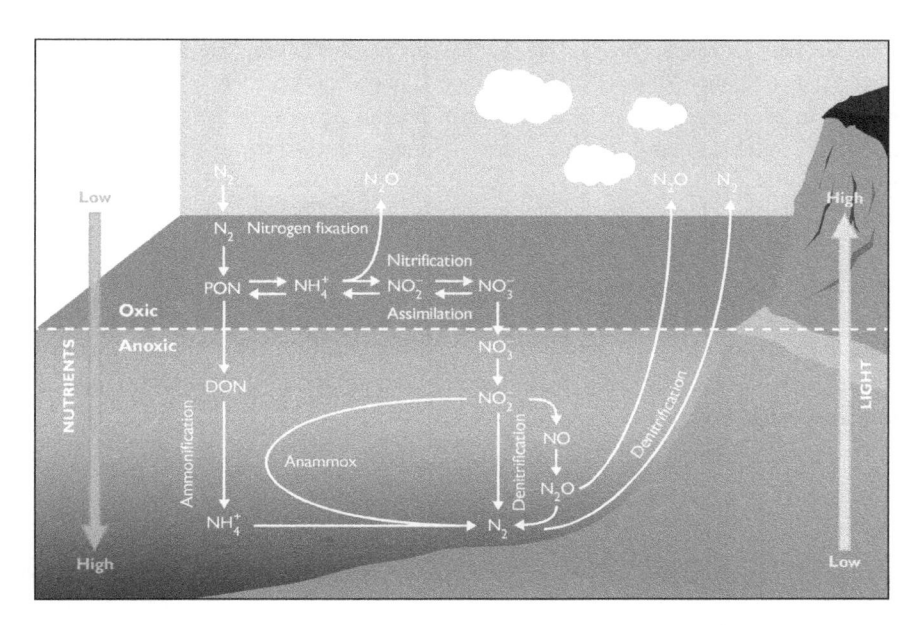

Figure 3. The marine N cycle, including losses of ammonium and nitrite as N_2: PON, particulate organic nitrogen, including phytoplankton; DON, dissolved organic nitrogen (modified from Arrigo, 2005).

microbes, is an important mechanism of nitrogen loss to the atmosphere and thus central to the nitrogen budget (Ward, 2013) (Figure 3). Specifically, in these environments, anaerobic remineralization of organic matter by denitrification and anaerobic ammonium oxidation (anammox) [the latter using nitrite (NO_2^-)] lead to a loss of bioavailable nitrogen through the formation of N_2 (Arrigo, 2005) as well as the production of N_2O (Codispoti et al., 2001) (Figure 3). Progressive deoxygenation of OMZ is expected to increase the volume of water where denitrification and anammox occur as well as increase the production of N_2O from denitrification, leading to increased nitrogen loss from the oceans (Bristow et al., 2017). However, it is generally assumed that this would only result in very short-term imbalances given that an increase in denitrification should cause an increase in N_2 fixation to balance it out (Gruber and Galloway, 2008). However, the direction and magnitude of change in the N_2O budget and air–sea N_2O flux remain unclear as increased stratification could decrease the amount of N_2O that reaches the surface ocean and is lost to the atmosphere (Martinez-Rey et al., 2015; Breitburg et al., 2018).

Another impact of N_2O is that, when it reaches the stratosphere, it destroys ozone (O_3), which acts as a protective layer to prevent UV radiation reaching the surface of Earth. In the process, a large proportion of N_2O reaching the stratosphere is actually destroyed in a reaction producing N_2:

$$N_2O \Rightarrow N_2 + O$$
$$N_2O + O \Rightarrow 2NO$$

Based on the work by Crutzen (1970) and Johnston (1971), nitrogen oxides ($NO_x = NO + NO_2$) catalytically destroy ozone in the stratosphere *via* the following reactions (Ravishankara et al., 2009):

$$NO + O_3 \Rightarrow NO_2 + O_2$$
$$O + NO_2 \Rightarrow NO + O_2$$
$$\text{net}: O + O_3 \Rightarrow 2O_2$$

By 2008, anthropogenic N_2O was the most significant ozone-destroying compound being emitted, and it is expected that N_2O will remain the most important anthropogenic O_3-destroying compound in the future (Portmann et al., 2012).

Biogeochemical Impacts of Deoxygenation in Sediments

Analysis of global benthic data indicates that denitrification occurring in sediments in OMZs and in sediments with high-nutrient and low oxygen levels removes around three times more nitrogen per unit of carbon sedimented than sediments underlying highly oxygenated water and accounts for ~10% (i.e., 15 Tg year^{-1}) of global benthic denitrification (Somes et al., 2013; Breitburg et al., 2018). Interestingly, anoxic conditions generally increase the release of phosphorus (Ingall and Jahnke, 1994) and iron (Scholz et al., 2014) from sediments, which have the potential to stimulate productivity if they reach the euphotic zone, such as above the OMZs in regions with coastal upwelling and the surface layer of coastal waters (Breitburg et al., 2018). This positive feedback has been observed in surface waters of fjords and estuaries when anoxia occurs and, as a result, dissolved inorganic phosphorus increases and thus phytoplankton productivity is stimulated in sunlit waters at the sea surface (Conley et al., 2007).

Estimates of Nitrogen Loss from OMZs

Loss of bioavailable nitrogen from the open ocean is estimated to be 65–80 Tg year^{-1} from the water column and 130–270 Tg year^{-1} from the sediments (Somes et al., 2013). OMZs appear to be responsible for ~30–50% of nitrogen loss from the world's oceans, or 16–27% from land and oceans combined (Codispoti et al., 2001; Gruber, 2004, 2008). It has been suggested that anammox in OMZ accounts for 10–15% of the loss of fixed N from the world's oceans (Arrigo, 2005) and that anammox will increase with the growth of OMZs (Hutchins and Fu, 2017). Also, OMZs limit atmospheric CO_2 sequestration by the ocean directly as CO_2 increases with remineralization (Paulmier et al., 2011) or indirectly through the decrease of total primary production by nitrogen loss (Falkowski, 1997).

Spreading of Dead Zones

Dead zones are hypoxic water bodies in the world's oceans driven primarily by eutrophication, caused by excessive nutrient pollution from human activities, depleting the oxygen required by marine life in bottom and near-bottom waters. These zones represent more than 245,000 square kilometers in the coastal oceans and have spread exponentially since the 1960s (Diaz and Rosenberg, 2008) largely because of warming and also because over the past half century, human use of nitrogen and phosphorus has exceeded the historical levels severalfold (Bouwman et al., 2013). Dead zones have now been reported from more than 400 systems, including highly productive continental seas, such as the Baltic, Kattegat, Black Sea, Gulf of Mexico, and East China Sea, all of which are major fishery areas (Diaz and Rosenberg, 2008). There are four types of dead zones — seasonal, permanent, temporary, and those governed by diel cycles. Approximately, 50% of the hypoxic zones worldwide are seasonal, usually lasting from a few weeks to several months. In seasonal hypoxic zones, oxygen depletion occurs in spring and summer following an increase in phytoplankton productivity as a result of nutrient enrichment and increase in light irradiance. Only ~8% of hypoxic zones are permanent, lasting throughout the year (Diaz and Rosenberg, 2008).

Examples of Physiological Effects of Ocean Deoxygenation

Most marine organisms (except those living in highly stable environments such as the deep sea) live in a dynamic environment where several parameters important to the fitness of that organism can change rapidly and simultaneously. All obligate aerobic organisms have constraints to the degree and duration of oxygen limitation/depletion, but generally, low-oxygen conditions lower survival, growth and reproduction (Diaz and Rosenberg, 2008; Sokolova, 2013). Some physiological adaptations to low-oxygen habitats include adjustments in ventilation rate, cardiac activity, hemoglobin content, O_2 binding and remodeled gill morphology to increase lamellar surface area in fish (Richards, 2011), and enhanced surface area:volume ratios to promote O_2 diffusion (Levin, 2003). In fish, as in other marine vertebrates, hemoglobin is involved in O_2 transport from respiratory organs to tissues. The affinity of blood for oxygen is the primary factor governing hypoxia tolerance in fish (Farrell and Richards, 2009; Mandic et al., 2009; Wells, 2009), and it is generally quantified as P_{50}, which is the partial pressure of oxygen at which blood is 50% saturated. Animals carrying blood with high affinity for oxygen have a low P_{50}, whereas those with low P_{50} blood are more tolerant to hypoxia.

Many marine animals depend on respiratory proteins (e.g., hemoglobin) to bind oxygen at respiratory surfaces (e.g., gills) to transport oxygen to the tissues and the CO_2 waste and protons to the gills for removal. In the majority of fish species, the oxygen affinity of hemoglobin is governed by protons, a phenomenon known as the Bohr effect. The oxygen transport function of respiratory proteins like hemoglobin is typically dependent on the blood pH and can be expressed as the Bohr coefficient ($\Delta \log P_{50} \Delta pH^{-1}$). Protons produced upon release of CO_2 decrease the affinity of respiratory proteins for O_2 (Hb–O_2+H$^+$ ⇔ Hb–H$^+$+O_2) (where Hb is hemoglobin). Therefore, the decrease in seawater pH caused by OA will reduce the effectiveness of O_2 uptake at the gills (Bridges and Morris, 1989; Seibel and Walsh, 2003). These acclimations come at an energetic cost and it is expected that hypoxia will lead to reduced feeding, growth, reproduction, and survival in some organisms, although with highly variable responses among species.

The reaction of oxygen binding to hemoglobin can be either exothermic or endothermic although in a few species heat is neither absorbed nor produced (Hochachka and Somero, 2002). In exothermic reactions, warming decreases oxygen affinity (i.e., increases P_{50}) because heat is produced when oxygen binds to hemoglobin. In endothermic reactions, warming increases oxygen affinity (i.e., decreases P_{50}) because heat is absorbed when oxygen binds to hemoglobin.

Hypoxia can also compromise vision in animals as oxygen demand is high in visual tissues. Specifically, photoreceptors (light-sensitive visual cells) as well as brain cells (neurons) have high requirement of oxygen for oxidative metabolism (Wong-Riley, 2010; McCormick and Levin, 2017), and consequently hypoxia can cause cell damage (McCormick and Levin, 2017). This is particularly relevant when oxygen decline occurs fast and under strong vertical gradients in oxygen concentrations, common to many parts of the ocean. The physiological effects of hypoxia on photoreceptor function and light sensitivity, in addition to morphological changes that may occur throughout ontogeny, have the potential to alter visual behavior and, subsequently, the ecology of marine animals, particularly fish, cephalopods and arthropods, which possess "fast" vision (McCormick and Levin, 2017).

Additionally, the negative effects of hypoxia can be transmitted across generations via epigenetic inheritance and can trigger changes expressed in future generations, even when the offspring has not been exposed to hypoxia (Wang *et al.*, 2016). Also, recurring exposure to low oxygen can alter immune responses, increase disease, and reduce growth (Keppel *et al.*, 2015). These transgenerational and epigenetic impacts of deoxygenation suggest that hypoxia might pose a dramatic and long-lasting threat to the sustainability of fish populations (Wang *et al.*, 2016).

Impacts of Hypoxia on Animal Behavior

The expansion of low-oxygen areas in the oceans is causing impairment of basic physiological functions as well as alteration of behavior in marine animals. In both coastal and open ocean systems, the spatial distribution of organisms and their preys is critically dependent upon oxygen gradients (Gilly *et al.*, 2013). Organisms that inhabit low-oxygen habitats have, over time, developed physiological and behavioral adaptations to take up, transport, and store oxygen to maintain aerobic metabolism (Seibel, 2011). However, the tolerance and behavior of organisms in response to changing oxygen vary between and within taxonomic groups (Vaquer-Sunyer and Duarte, 2008), thus causing alterations in predator–prey interactions and the structure of marine food webs (Breitburg *et al.*, 2018).

Fish display many different physiological responses to hypoxia (Zhu *et al.*, 2013) although a typical adaptation to chronic hypoxia involves a decrease in scope for activity by decrease of maximal metabolic rates

(Seebacher and Krause, 2017). However, while adaptations to hypoxia may cause fish to decrease activity to conserve energy, in some cases fish actually become more active to reach environments with higher O_2 levels (Seebacher and Krause, 2017; Domenici et al., 2017). The individual physiological responses to hypoxia caused by changes in access to oxygenated water are often translated into alterations in animal behavior including group coordination, maximum group sizes and population structure (Seebacher and Krause, 2017). The effect of hypoxia generates trade-offs that affect schooling positions and school structure, both important in the survival of fish populations (Domenici et al., 2017; Seebacher and Krause, 2017). Interestingly, observations in the field have also revealed that large schools of fish can worsen hypoxic conditions, with potential consequences for school structure and size (Domenici et al., 2017).

Ecological Impacts of Hypoxia

As hypoxia increases in the oceans due to the continued spread of coastal eutrophication and global warming, organisms that require oxygen must adapt. However, thresholds for hypoxia vary greatly across marine organisms (Vaquer-Sunyer and Duarte, 2008) with benthic organisms being particularly vulnerable to coastal hypoxia because their habitats are distant from atmospheric oxygen supply and also because coastal sediments are usually depleted in oxygen relative to the overlying water column.

Hypoxia has a direct effect on marine fauna through the exclusion of non-tolerant species and, in some cases, it triggers mortality events in "dead zones", often observed in metazoans, for example, fish and crustaceans (Renaud, 1986; Rabalais et al., 2002). Sublethal stresses include reduced biodiversity, decreases in growth and reproduction, physiological stress, forced migration, habitat compression, increased vulnerability to predation, and disruption of life cycles (Service, 2004). Benthic organisms exhibit signs of stress and eventually die when hypoxia is strong and persistent. Because of these highly restricted ecological niches, benthic communities in areas exposed to seasonally severe hypoxia are generally less diverse, less abundant, and less productive than benthic habitats with well-oxygenated waters (Baustian et al., 2009).

Hypoxia can also change the spatial distribution and the community structure of fish through mortality during extreme events (Domenici et al., 2017). In less extreme conditions, hypoxia can cause a decrease in growth in fish, which can have detrimental effects on reproduction (Taylor and

Miller, 2001). For example, severe reproductive impairment in Atlantic croaker has been found to occur over large coastal regions when their populations are exposed to seasonal hypoxia (Thomas and Rahman, 2012). Like with many other stressors, hypoxia can cause energetic trade-offs in marine organisms, resulting in shifts in behavior as well as ecological fitness.

Predator–prey interactions

In addition to changes in the physiology, morphology, behavior and ecology of marine organisms in response to hypoxia, there are also indirect effects of hypoxia, such as the interaction between predator and prey with different oxygen tolerances (Robb and Abrahams, 2003), which is important in understanding and managing marine ecosystems. An example is mobile demersal (living on or near the bottom) animals, which can swim to shallow waters when the lower part of the water column is hypoxic, but being closer to the surface exposes them to increased predation.

Hypoxia can also impair visual function in marine animals and oxygen stress may trigger a change in the light environment as a result of habitat compression, or cause a physiological change in light requirements (Wong-Riley, 2010; McCormick and Levin, 2017). Organisms exposed to chronically low-oxygen environments, such as those living in oxygen-minimum zones may already employ adaptations to sustain visual function, whereas species exposed to short-term oxygen-limiting conditions may be less tolerant to these conditions and at risk for a decline in visual function (McCormick and Levin, 2017). Marine birds and mammals that rely on fish and crustaceans, such as herons or otters, can be affected by a decline in fisheries caused by hypoxia, and thus hypoxic areas may be more vulnerable to overfishing, among other stressors. These impacts could alter the distribution and behavior of visual marine animals and increase their vulnerability to predators or impact their feeding success.

Effect of hypoxia on habitat compression

The projected warmer temperatures and lower oxygen concentrations in the oceans cause "habitat compression" (Prince and Goodyear, 2007; Stramma et al., 2012), which changes ecosystem structure and species' ecological interactions (Chu and Tunnicliffe, 2015). Specifically, climate change is projected to change P_{50} depths in many areas and alter the depth ranges occupied by fish (Figure 4). For example, shoaling of P_{50} depths in some tuna species is expected to result in a compression of their vertical habitat,

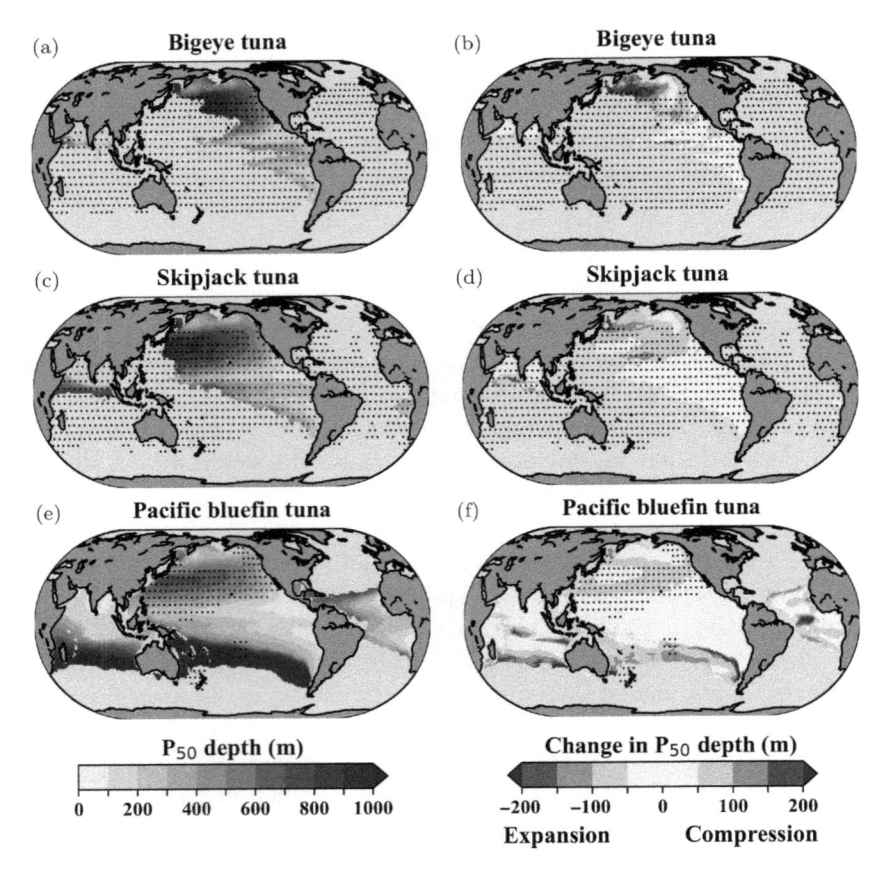

Figure 4. P_{50} depths and the projections of changes in P_{50} depths as a result of climate change (from Mislan et al., 2017). Projections are based on different models (see Mislan et al., 2017 for details). Grey indicates locations where there are no P_{50} depths. (a), (c), and (e): Present-day P_{50} depths of bigeye, skipjack, and Pacific bluefin tuna calculated using data from World Ocean Atlas 2009. (b), (d), and (f): Average projected changes in P_{50} depths for bigeye, skipjack, and Pacific bluefin tunas from the six Earth System Models included in the Climate Model Intercomparison Project 5. Expansion occurs in locations where P_{50} depths are deeper in the future. Compression occurs in locations where P_{50} depths are shallower in the future. The stippling indicates known habitat for each tuna species (IUCN, 2011, 2014).

in the Northwest Pacific Ocean, experiencing the most pronounced shoaling (more than 200 m) (Figure 4) although some areas appear to result in deeper P_{50} depths, indicating an expansion of the vertical habitat (Mislan et al., 2017).

Shoaling of the tropical OMZ decreases the depth range of tropical pelagic fishes such as marlins, sailfish, and tuna by compressing their habitat into a narrow surface layer (Prince and Goodyear, 2006; Stramma et al., 2010). Hypoxia at depth limits the vertical movements of some marine animals. For example, some tuna species like bigeye tuna remain near the ocean surface at night and move to deeper layers just above the oxygen-minimum layers during the day (Schaefer and Fuller, 2010; Mislan et al., 2017). However, while in deep waters, bigeye tunas repetitively return to shallower depths because they cannot remain in hypoxic conditions for too long (Schaefer and Fuller, 2010). On the other hand, other types of tuna (e.g., yellowfin and skipjack) preferentially live in surface waters both during night and day although they can swim to deeper waters during the day to feed if food is not available at the surface (Schaefer et al., 2009; Mislan et al., 2017). Like bigeye tuna, yellowfin and skipjack can only remain in deeper waters for short periods (Schaefer et al., 2009). Restriction of these fishes in shallower waters could restrict their food availability and also make fish more vulnerable to fishing and alter the predator–prey dynamics (Prince and Goodyear, 2006).

Key Points

- Warming, eutrophication and reduced ventilation are causing major deoxygenation in many parts of the ocean (Shepherd et al., 2017).
- Hypoxia is generally harmful and, in some cases, lethal to marine life.
- Eutrophication of coastal systems and enclosed basins (e.g., Baltic Sea, Black Sea) has a major contribution to hypoxia through excessive production of organic matter, promoting growth of bacteria and increasing the oxygen demand of the ecosystem.
- It has been suggested that anammox will increase with the expansion of OMZs (Arrigo, 2005).
- The expansion of OMZs might limit atmospheric CO_2 sequestration by the ocean via remineralization, which produces CO_2 (Paulmier et al., 2011) or through limitation of primary production by nitrogen loss (Falkowski, 1997).
- A major factor affecting tolerance for hypoxia is the affinity of blood for oxygen. Neurons have high requirement of oxygen for oxidative metabolism, and oxygen demand is high in visual tissues, which can impact behavior.

- Approximately, 50% of the hypoxic zones worldwide are seasonal, usually lasting from a few weeks to several months while ~8% of hypoxic zones are permanent (Diaz and Rosenberg, 2008).
- Habitat compression in pelagic ecosystems forces organisms to quickly acclimate to new environments, increases their vulnerability to predation, and disrupts life cycles. Benthic organisms can, in some cases, die when hypoxia is strong and persistent.

Questions

1. What process(es) in the water column contribute to the release of nitrous oxide?
2. Where does deoxygenation occur in the water column and what processes cause it?
3. What biologically mediated processes lead to the loss of bioavailable nitrogen through the formation of N_2?
4. Give examples of physiological adaptations to low-oxygen habitats.
5. Is the P_{50} (the partial pressure of oxygen at which blood is 50% saturated) higher or lower in animals tolerant to hypoxia? Why?
6. Explain how eutrophication affects oxygen levels in seawater.
7. How does the expansion of OMZs affect CO_2 levels through the water column?
8. How do you expect the $CO_2:O_2$ ratio change at the ocean surface with climate change (consider warming, ocean acidification and deoxygenation)?
9. Give examples of some economic impacts that could occur due to the increase in dead zones.
10. In what season would you expect hypoxia to be more pronounced and why?

References

Arrigo KR (2005) Marine microorganisms and global nutrient cycles. *Nature* **437**, 349–355.

Baustian MM, Craig JK, Rabalais NN (2009) Effects of summer 2003 hypoxia on macrobenthos and Atlantic croaker foraging selectivity in the northern Gulf of Mexico. *Journal of Experimental Marine Biology and Ecology* **381**, S31–S37.

Bange SW, Wajih S, Naqvi A, Codispoti LA (2005) The nitrogen cycle in the Arabian Sea. *Progress in Oceanography* **65**, 145–158.

Bouwman L, Goldewijk K, Van Der Hoek KW, Beusen AHW, Van Vuuren DP, Willems J, Rufino MC, Stehfest E (2013) Exploring global changes in nitrogen and phosphorus cycles in agriculture induced by livestock production over the 1900–2050 period. *Proceedings of the National Academy of Sciences USA* **110**, 20882–20887.

Breitburg D, Levin LA, Oschlies A, Grégoire M, Chavez FP, Conley DJ, Garçon V, Gilbert D, Gutierrez D, Isensee K, Jacinto GS, Limburg KE, Montes I, Naqvi SWA, Pitcher GC, Rabalais NN, Roman MR, Rose KA, Seibel BA, Telszewski M, Yasuhara M, Zhang J (2018) Declining oxygen in the global ocean and coastal waters. *Science* **359**, DOI: 10.1126/science.aam7240.

Bridges CR, Morris S (1989) Respiratory pigments: Interactions between oxygen and carbon dioxide transport. *Canadian Journal of Zoology* **67**, 2971–2985.

Bristow LA, Callbeck CM, Larsen M, Altabet MA, Dekaezemacker J, Forth M, Gauns M, Glud RN, Kuypers MMM, Lavik G, Milucka J, Naqvi SWA, Pratihary A, Revsbech NP, Thamdrup B, Treusch AH, Canfield DE (2017) N_2 production rates limited by nitrite availability in the Bay of Bengal oxygen minimum zone. *Nature Geoscience* **10**, 24–29.

Chu JWF, Tunnicliffe V (2015) Oxygen limitations on marine animal distributions and the collapse of epibenthic community structure during shoaling hypoxia. *Global Change Biology* **21**, 2989–3004.

Cline JD, Richards FA (1972) Oxygen-deficient conditions and nitrate reduction in the eastern tropical North Pacific. *Limnology and Oceanography* **17**, 885–900.

Codispoti LA, Brandes JA, Christensen JP, Devol AH, Naqvi SWA, Paerl HW, Yoshinari T (2001) The oceanic fixed nitrogen and nitrous oxide budgets: moving targets as we enter the anthropocene? *Science Marine* **65**, 85–105.

Conley DJ, Carstensen J, Ærtebjerg G, Christensen PB, Dalsgaard T, Hansen JLS, Josefson AB (2007) Long-term changes and impacts of hypoxia in Danish coastal waters. *Ecological Applications* **17(5)**, S165–S184.

Crutzen PJ (1970) The influence of nitrogen oxides on atmosphere ozone content. *Quarterly Journal of the Royal Meteorological Society* **96**, 320–325.

Deutsch C, Emerson S, Thompson L (2005) Fingerprints of climate change in North Pacific oxygen. *Geophysical Research Letters* **32**, L16604, doi:10.1029/2005GL023190.

Deutsch C, Brix H, Ito T, Frenzel H, Thompson L (2011) Climate-forced variability of ocean hypoxia. *Science* **333**, 336–339.

Diaz RJ (2001) Overview of hypoxia around the world. *Journal of Environmental Quality* **30**, 275–281.

Diaz RJ, Rosenberg R (2008) Spreading dead zones and consequences for marine ecosystems. *Science* **321**, 926–929.

Domenici P, Steffensen JF, Marras S (2017) The effect of hypoxia on fish schooling. *Philosophical Transactions of the Royal Society B* **372**, 20160236.

Falkowski PG (1997) Evolution of the nitrogen cycle and its influence on the biological sequestration of CO_2 in the ocean. *Nature* **387**, 272–275.

Farrell AP, Richards JG (2009) Defining hypoxia: An integrative synthesis of the responses of fish to hypoxia. *Fish Physiology* **27**, 487–503.

Gilly WF, Beman JM, Litvin SY, Robison BH (2013) Oceanographic and biological effects of shoaling of the oxygen minimum zone. *Annual Review of Marine Science* **5**, 393–420.

Gruber N (2004) The dynamics of the marine nitrogen cycle and atmospheric CO_2. In: Oguz T., Follows M., (eds.) *Carbon Climate Interactions*. Kluwer, Dordrecht, pp. 97–148.

Gruber N (2008) The Marine Nitrogen Cycle: Overview and Challenges. In: Capone D. G., Bronk D. A., Mulholland M. R., Carpenter E. J. (eds.) *Nitrogen in the Marine Environment*, Elsevier, MA, USA.

Gruber N, Galloway JN (2008) An Earth-system perspective of the global nitrogen cycle. *Nature* **451**, 293–296.

Hochachka PW, Somero GN (2002) *Biochemical Adaptation: Mechanism and Process in Physiological Evolution*. New York, Oxford University Press, p. 466.

Hutchins DA, Fu F (2017) Microorganisms and ocean global change. *Nature Microbiology* **2**, 170588.

Ingall E, Jahnke R (1994) Evidence for enhanced phosphorus regeneration from marine sediments overlain by oxygen depleted waters. *Geochimicaet Cosmochimica Acta* **58**, 2571–2575.

IUCN (2011) *Thunnus maccoyii, Katsuwonus pelamis, Thunnus alalunga, Thunnus obesus, Thunnus thynnus, Thunnus albacares, Thunnus atlanticus, Thunnus tonggol* spatial data, Version 2016-2. The IUCN Red List of Threatened Species.

IUCN (2014) *Thunnus orientalis* spatial data, Version 2016-2. The IUCN Red List of Threatened Species.

Johnston H (1971) Reduction of stratospheric ozone by nitrogen oxide catalysts from supersonic transport exhaust. *Science* **173**, 517–522.

Karstensen J, Stramma L, Visbeck M (2008) Oxygen minimum zones in the eastern tropical Atlantic and Pacific oceans. *Progress in Oceanography* **77**, 331–350.

Keeling RF (2016) Atmospheric oxygen data for La Jolla, California. See http://scrippso2.ucsd.edu/index.

Keeling RF, Garcia H (2002) The change in oceanic O_2 inventory associated with recent global warming. *Proceedings of the National Academy of Sciences* **99**, 7848–7853.

Keeling RF, Körtzinger A, Gruber N (2010) Ocean deoxygenation in a warming world. *Annual Review of Marine Science* **2**, 199–229.

Keppel AG, Breitburg DL, Wikfors GH, Burrell RB, Clark VM (2015) Effects of co-varying diel-cycling hypoxia and pH on disease susceptibility in the eastern oyster *Crassostrea virginica*. *Marine Ecology Progress Series* **538**, 169–183.

Levin L (2003) Oxygen minimum zone benthos: Adaptation and community response to hypoxia. *Oceanography and Marine Biology* **41**, 1–45.

Mandic M, Todgham AM, Richards JG (2009) Mechanisms and evolution of hypoxia tolerance in fish. *Proceedings of the Royal Society B* **276**, 735–744.

Martinez-Rey J, Bopp L, Gehlen M, Tagliabue A, Gruber N (2015) Projections of oceanic N_2O emissions in the 21st century using the IPSL Earth system model. *Biogeosciences* **12**, 4133–4148.

McCormick LR, Levin LA (2017) Physiological and ecological implications of ocean deoxygenation for vision in marine organisms. *Philosophical Transactions of the Royal Society A* **375**, 20160322.

Mislan KAS, Deutsch CA, Brill RW, Dunne JP, Sarmiento JL (2017) Projections of climate-driven changes in tuna vertical habitat based on species-specific differences in blood oxygen affinity. *Global Change Biology* **23**, 4019–4028.

Paulmier A, Ruiz-Pino D, Garcon V (2008) The oxygen minimum zone (OMZ) off Chile as intense source of CO_2 and N_2O. *Continental Shelf Research* **28**, 2746–2756.

Paulmier A, Ruiz-Pino D, Garon V (2011) CO_2 maximum in the oxygen minimum zone (OMZ). *Biogeosciences* **8**, 239–252.

Portmann RW, Daniel JS, Ravishankara AR (2012) Stratospheric ozone depletion due to nitrous oxide: Influences of other gases. *Philosophical Transactions of the Royal Society B* **367**, 1256–1264.

Prince ED, Goodyear CP (2006) Hypoxia-based habitat compression of tropical pelagic fishes. *Fisheries Oceanography* **15**, 451–464.

Prince ED, Goodyear CP (2007) Consequences of ocean scale hypoxia constrained habitat for tropical pelagic fishes. *Gulf and Caribbean Research* **19**, 17–20.

Rabalais NN, Turner RE, Wiseman WJ (2002) Gulf of Mexico hypoxia, aka "The dead zone." *Annual Review of Ecology and Systematics* **33**, 235–263.

Ravishankara AR, Daniel JS, Portmann RW (2009) Nitrous oxide (N_2O): The dominant ozone-depleting substance emitted in the 21st century. *Science* **326**, 123–125.

Renaud M (1986) Hypoxia in Louisiana coastal waters during 1983: Implications for fisheries. *Fisheries Bulletin* **84**, 19–26.

Richards JG (2011) Metabolic and molecular responses of fish to hypoxia. *Fish Physiology* **27**, 443–485.

Robb T, Abrahams MV (2003) Variation in tolerance to hypoxia in a predator and prey species: An ecological advantage of being small? *Journal of Fish Biology* **62**, 1067–1081.

Santoro AE, Buchwald C, McIlvin MR, Casciotti KL (2011) Isotopic signature of N_2O produced by marine ammonia-oxidizing archaea. *Science* **333**, 1282–1285.

Sarmiento JL, Hughes TMC, Stouffer RJ, Manabe S. (1998) Simulated response of the ocean carbon cycle to anthropogenic climate warming. *Nature* **393**, 245–49.

Schaefer KM, Fuller DW (2010) Vertical movements, behavior, and habitat of bigeye tuna (*Thunnus obesus*) in the equatorial eastern Pacific Ocean, ascertained from archival tag data. *Marine Biology* **157**, 2625–2642.

Schaefer KM, Fuller DW, Block BA (2009) Vertical movements and habitat utilization of skipjack (*Katsuwonus pelamis*), yellowfin (*Thunnus albacares*), and bigeye (*Thunnus obesus*) tunas in the Equatorial Eastern Pacific Ocean, ascertained through archival tag data. In: Nielsen, J. L., Arrizabalaga, H.,

Fragoso, N., Hobday, A., Lutcavage, M., Sibert, J. (eds.), *Tagging and Tracking of Marine Animals With Electronic Devices* New York, Springer, pp. 121–144.

Schmidtko S, Stramma L, Visbeck M (2017) Decline in global oceanic oxygen content during the past five decades. *Nature* **542**, 335–339.

Scholz F, McManus J, Mix AC, Hensen C, Schneider RR (2014) The impact of ocean deoxygenation on iron release from continental margin sediments. *Nature Geoscience* **7**, 433–437.

Seebacher F, Krause J (2017) Physiological mechanisms underlying animal social behaviour. *Philosophical Transactiosn of the Royal Society B* **372,** 20160231.

Seibel BA (2011) Critical oxygen levels and metabolic suppression in oceanic oxygen minimum zones. *Journal of Experimental Biology* **214**, 326–336.

Seibel BA, Walsh PJ (2003) Biological impacts of deep-sea carbon dioxide injection inferred from indices of physiological performance. *The Journal of Experimental Biology* **206**, 641–650.

Service RF (2004) New dead zone off Oregon coast hints at sea change in current, *Science* **305**, 1099.

Shepherd JG, Brewer PG, Oschlies A, Watson AJ (2017) Ocean ventilation and deoxygenation in a warming world: Introduction and overview. *Philosophical Transactions of the Royal Society A* **375**, 20170240.http://dx.doi.org/10.1098/rsta.2017.0240.

Sokolova IM (2013) Energy-limited tolerance to stress as a conceptual framework to integrate the effects of multiple stressors. *Integrative and Comparative Biology* **53**, 597–608.

Somes CJ, Oschlies A, Schmittner A (2013) Isotopic constraints on the pre-industrial oceanic nitrogen budget. *Biogeosciences* **10**, 5889–5910.

Stendardo I, Gruber N (2012) Oxygen trends over five decades in the North Atlantic. *Journal of Geophysical Research* **117**, C11004.

Stramma L, Johnson GC, Sprintall J, Mohrholz V (2008) Expanding oxygen minimum zones in the tropical oceans, *Science* **230**, 655–658.

Stramma L, Schmidtko S, Levin LA, Johnson GC (2010) Ocean oxygen minima expansions and their biological impacts. *Deep-Sea Research I* **57**, 587–595.

Stramma L, Prince ED, Schmidtko S, Luo J, Hoolihan JP, Visbeck M, Wallace DWR, Brandt P, Körtzinger A (2012) Expansion of oxygen minimum zones may reduce available habitat for tropical pelagic fishes. *Nature Climate Change* **2**, 33–37.

Taylor JC, Miller JM (2001) Physiological performance of juvenile southern flounder, *Paralichthys lethostigma* (Jordan and Gilbert 1884), in chronic and episodic hypoxia. *Journal of Experimental Marine Biology and Ecology* **258**, 195–214.

Thomas P, Rahman MS (2012) Extensive reproductive disruption, ovarian masculinization and aromatase suppression in Atlantic croaker in the northern Gulf of Mexico hypoxic zone. *Proceedings of the Royal Society B* **279**, 28–38.

Vaquer-Sunyer R, Duarte CM (2008) Thresholds of hypoxia for marine biodiversity. *Proceedings of the National Academy of Sciences* **105**, 15452–15457.

Wang SY, Lau K, Lai K-P, Zhang J-W, Tse AC-K, Li J-W, Tong Y, Chan T-F, Wong CK-C, Chiu JM-Y, Au DW-T, Wong AS-T, Kong RY-C, Wu RS-S (2016) Hypoxia causes transgenerational impairments in reproduction of fish. *Nature Communications* **7**, 12114.

Ward BB (2013) How nitrogen is lost. *Science* **341**, 352–353.

Wells RMG (2009) Chapter 6 Blood-gas transport and hemoglobin function: Adaptations for functional and environmental hypoxia. *Fish Physiology* **27**, 255–299.

Wong-Riley M (2010) Energy metabolism of the visual system. *Eye Brain* **2**, 99–116.

Wright JJ, Konwar KM, Hallam SJ (2012) Microbial ecology of expanding oxygen minimum zones. *Nature Reviews Microbiology* **10**, 381–394.

Wyrtki K (1962) The oxygen minima in relation to ocean circulation. *Deep-Sea Research* **9**, 11–23.

Zhu C-D, Wang Z-H, Yan B (2013) Strategies for hypoxia adaptation in fish species: A review. *Journal of Comparative Physiology B* **183**, 1005–1013.

Chapter 5

Plastic Pollution
Excessive Plastic Production and Consumption Is Filling Up the Oceans

"One of the most sinister features of DDT and related chemicals is the way they are passed on from one organism to another through all the links of the food chains. For example, fields of alfalfa are dusted with DDT; meal is later prepared from the alfalfa and fed to hens; the hens lay eggs which contain DDT. Or the hay, containing residues of 7 to 8 parts per million, may be fed to cows. The DDT will turn up in the milk in the amount of about 3 parts per million, but in butter made from this milk the concentration may run to 65 parts per million. Through such a process of transfer, what started out as a very small amount of DDT may end as a heavy concentration."

— From chapter "Elixirs of Death" in *Silent Spring* by Rachel Carson (1962), on the transfer of DDT through the food chain, comparable to the bioaccumulation of plastic and its additives through the marine food web.

Introduction

We live in the "Plasticene", the age of massive production of plastic by humans, with an annual release of more than 300 million tonnes (Mt), which approaches the total weight of the human population produced in plastic every year (Worm et al., 2017; ten Brink et al., 2018). Plastics are synthetic organic polymers resulting from the polymerization of monomers extracted from oil or gas (Thompson *et al.*, 2009). Since the birth of the first synthesized plastic in 1885 by Alexander Parker, the invention of the first modern plastic (bakelite) by Leo Hendrik Baekeland in 1907, and the commercial growth of plastic production during World War II (SPI, 2015), manufacturing techniques have been optimized for mass production of inexpensive, versatile, lightweight, durable, inert, and corrosion-resistant plastics.

The first reports of plastic waste in the oceans in the early 1970s (Colton and Knapp, 1974) drew little interest among the scientific community. Half a century later, plastic has become an indispensable material used for many applications in homes and industry, and the vast consumption of plastic has been generating massive waste that is causing environmental problems in the ocean (Figure 1). Both as visible litter such as the conspicuous accumulation of plastic debris and also as invisible particles, such as microfibers and nanoplastics, plastic pollution has reached a global dimension. Indeed, the global production of plastic has been increasing every year. In 2015, global plastic production rose to 322 Mt/year (Cressey, 2016; Worm *et al.*, 2017),

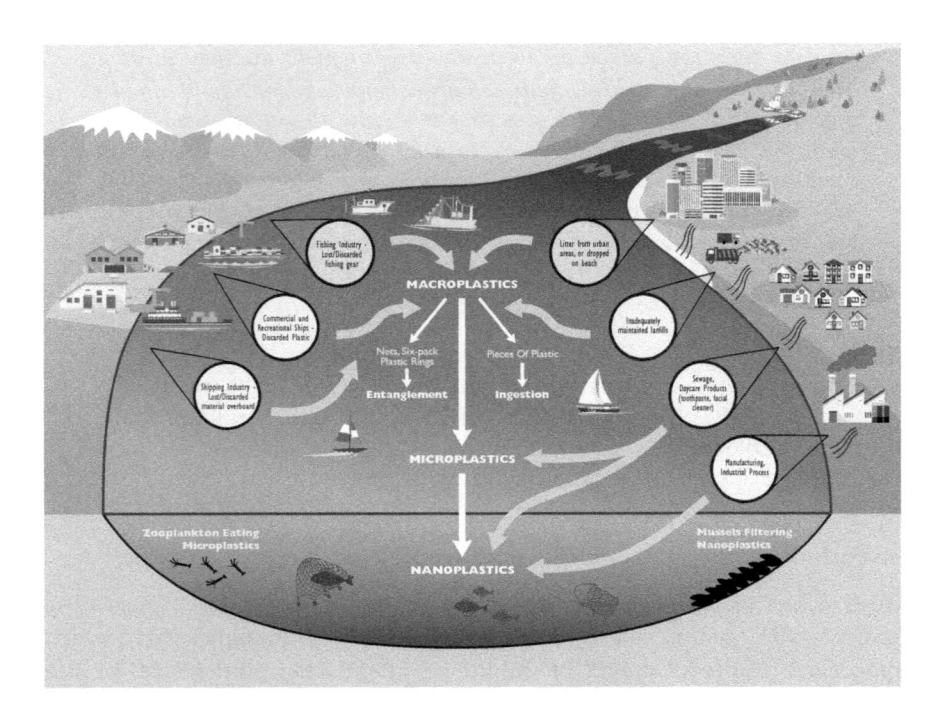

Figure 1. Sources of marine plastic pollution.

of which 5–15 Mt is thought to end up in the already troubled marine environment (ten Brink et al., 2018). The U.S. appears to be among the highest producer of plastic waste per person per day (Jambeck et al., 2015).

Plastic is the most abundant component of marine litter and its proportion varies between 60% and 80% of the total marine debris (Gregory and Ryan, 1997), sometimes accounting for up to 100% of floating waste (Galgani et al., 2015). Plastic pollution has reached every single corner of the ocean from polar to tropical ecosystems, both coastal and open oceans, and in surface and sediments (Barnes et al., 2009). The uniqueness of plastic lies in that while other human-produced waste usually degrades/decomposes or is corroded over short time periods, plastics decompose only slowly or not at all, and some can remain in seawater for very long periods of time (years, decades, centuries). This slow degradation of plastic is also seen on the seafloor, where 90% of litter caught in benthic trawls is plastic (Galgani et al., 2000; Ramirez-Llodra et al., 2013).

Notably, the "Great Pacific garbage patch", a region of the Central Pacific, where plastic debris accumulates has received much attention, but scientists still struggle to determine the mechanisms governing plastic accummulation in the oceans, its distribution, and the distinct effects of different types of plastic on marine ecosystems. Increasingly, more and more marine life appears to be harmed by plastic pollution through entanglement and ingestion of plastic debris as well as the "hitchhiking" on drifting plastic debris by "invader" species, and the absorption of polychlorinated biphenyls and other organic pollutants from ingested plastics (Derraik, 2002). Less explored impacts of plastic pollution include damage to subsistence fisheries (Nash, 1992), the formation of sheeting that blankets the biota of soft sediment, reef, and rocky substrata (Uneputty and Evans, 1997), and hypoxia and even anoxia induced by reduction of gas exchange between pore waters in sediments and overlying seawater caused by plastic pieces (Goldberg, 1997; Gregory and Andrady, 2003).

As the rapid production of plastics continues to grow, in contrast with the slow growth of plastic recycling rates (US EPA, 2014; *Plast. Eur.*, 2015), the amount of plastic pollution in the ocean is likely to continue to increase. Disconcertingly, plastic in the ocean is projected to triple by 2025 (Government Office for Science, 2018). This is worrying because although the extent of the impacts of plastic pollution is not yet well established, there is growing evidence of chemical pollution, disruption of the endocrine system of marine animals, damage to organs when plastic is ingested, and restriction of organism's movement via entanglement (Thompson, 2017). This chapter addresses the state of knowledge and consequences of plastic pollution in the oceans exploring the effects of plastic and its additives on marine organisms and ecosystems.

Sources of Plastic Pollution

The marine environment has become an "oceanfill" for waste produced by humans. The main source of marine plastic pollution globally is by far plastic waste produced on land, and the main cause appears to be poorly managed plastic waste (Jambeck et al., 2015). The origin of plastic waste entering the ocean is diverse and includes street litter, landfills, industrial waste, and production waste (Isensee and Valdes, 2015) (Figure 2). However, marine plastic litter also originates from activities at sea including those at commercial and recreational ships, fishing and aquaculture operations (rope, waste, fishing gear, nets), for example, when fishing gear or cargo is lost, abandoned or discarded (Figure 1), despite international legislation prohibiting the dumping of waste at sea [MARPOL International Convention for the Prevention of Pollution from Ships, 1988; Jones, 1995; Isensee and Valdes, 2015].

It has been estimated that between 4.4 and 12.7 million Mt of plastics enter the ocean every year as macroscopic litter and microplastic particles (Jambeck et al., 2015; Worm et al., 2017). Although comparable datasets are not available globally, based on 12,000 measurements of plastic abundance between 1971 and 2013 (Van Sebille et al., 2015), the global mass

Figure 2. Pictures depicting sources of plastic pollution that end up in the ocean in Vietnam, the fourth most polluted country worldwide (Jambeck et al., 2015). Plastic pollution at Mekong River, the tenth most polluted river (Schmidt et al., 2017) (left); unregulated landfill in Phú Quốc (middle), and typical small, scale recycling business in Vietnam (right). Images were kindly provided by Dr. Frauke Bagusche, who documented plastic pollution during her global voyages.

inventory of plastic collected by nets in the upper ocean varies between 93,000 to 236,000 Mt (Law, 2017). There are, however, large uncertainties in these types of estimates because the flux of waste entering the ocean from land (e.g., rivers, wind, tidal, and ocean wave transport) is actually not measured (Law, 2017) and its distribution is highly dynamic.

Types and Properties of Plastics

Plastics can be classified into thermoplastics and thermoset plastics. Thermoplastics contain ingredients aimed at improving the physical properties and performance of plastic (Alauddin *et al.*, 1995; Plastics Europe, 2016). Thermoplastics can be repeatedly melted when heated and hardened when cooled such that this material can be reheated, reshaped and frozen, and it is resistant to alteration. Also because thermoplastics are the most abundant type of plastic accounting for ∼85% of total plastic demand in practically all market sectors (The American Chemistry Council, 2015), they are an important source of non-biodegradable plastics to the environment. On the other hand, thermoset plastics or thermosets are solidified after being melted by heating and, unlike thermoplastics, thermosets are susceptible to alteration (Muller *et al.*, 2001).

The most common types of plastics include polyethylene (PE), polypropylene (PP), polystyrene (PS), polycarbonate (PC), polyvinylchloride (PVC), polyamide, polyethylene terephthalate (PET), polyurethane (PU), and polyvinyl alcohol (Plastics Europe, 2016; Kedzierski *et al.*, 2017, 2018; Worm *et al.*, 2017) (Table 1). These plastics made up >80% of the plastic use in Europe in 2016, and a large fraction of global plastic production (Plastics Europe, 2016). They differ in their chemical composition and environmental impacts. Most PPs and PEs are used to make pliable films and packaging materials as well as vehicle parts, pipes, and houseware; PVC and PU are used in construction and the transportation industries; PET is used for manufacturing textile fibers and drink bottles; PS is used for packaging (e.g., Styrofoam) and building insulation; PC is used in hard, transparent products (e.g., eyeglasses, clear roofing sheets) (Worm *et al.*, 2017).

During the transformation of plastic into a product, additives may be added to shape their physical characteristics, achieve desired plastic properties, and improve performance and appearance (Law, 2017). These plastic additives are called plasticizers, which are typically added to promote flexibility (phtalates and adipates), and include metals (antimony,

Table 1. Types of plastic commonly encountered in the marine environment [from American Plastics Council Year End Statistics (2004); Zheng et al. (2005); Andrady (2011)].

Type of plastic	Application
Polypropylene (PP)	Tanks, jugs, containers, rope, bottle caps, nets
High-density polyethylene (PE)	Bottles, milk/juice jugs, pipes, insulation
Polyvinyl chloride (PVC)	Films and packaging
Low-density polyethylene (LDPE/LLDPE)	Films, packaging, plastic bags, six-pack rings, bottles, netting, drinking straws
Polystyrene (PS)	Containers, plastic utensils, food containers
Foamed polystyrene	Floats, bait boxes, foam cups
Nylon	Netting and traps
Thermoplastic Polyester (PET)	Plastic bottles for beverages
Poly(vinyl chloride) (PVC)	Plastic film, bottles, cups
Thermosets and other plastics	Lamination, furniture, paper, clothing, molding
Cellulose Acetate (CA)	Cigarette filters
Total	

lead), antioxidants (phenolic and phosphite compounds), UV stabilizers (benzotriazole and titanium dioxide), fillers (calcium carbonate), dyes (titanium dioxide) and antimicrobial ingredients (e.g., Triclosan, silver) that can leach into the marine environment causing harm to marine organisms (Kedzierski et al., 2018). Also, plastic can adsorb toxic compounds such as heavy metals (e.g., cadmium and lead) discharged by industry (e.g., Holmes et al., 2012; Boucher et al., 2016), and organic molecules that can bind to the hydrophobic surface of plastics (Rios et al., 2007). These include dichlorodiphenyltrichloroethane (DDT) and polychlorinated biphenyls (PCBs), both linked directly or indirectly to cancer (e.g., Hiatt and Green Brody, 2018).

Size and Distribution of Plastic Through the Water Column

There are many studies reporting particle size information of microplastics in water and sediments (e.g., Hidalgo-Ruz et al., 2012) although there are large discrepancies in particle numbers as well as the shape of plastic (e.g., spheres, fibers, irregular pieces). Additionally, the size classes of plastic change because the ocean is a dynamic system where plastic is weathered over time and breaks down into smaller pieces (Law, 2017). For example, biofouling and exposure of plastic to UV radiation leads to photo-oxidation that decreases the plastic molecular weight, makes it more susceptible to shear or tensile stresses, and causes fracturing and fragmentation (Andrady 2015). This means that plastic pollution might become less conspicuous but the amount of plastic remains the same. This "invisible" plastic has different and perhaps more serious impacts on marine life and interacts with the base of the food chain (see effects of nano- and microplastics in "Plastic Degradation in the Ocean" sections, and in subsections: "Trophic transfer and bioaccumulation of plastic" and "Toxicity of plastics as endocrine-disrupting chemicals").

Plastic can be divided into a few categories based on size. Nanoplastics are the smallest plastic particles (<1 µm in diameter), and they display a colloidal[1] behavior (Gigault et al., 2018). Nanoplastics largely originate from manufacturing, industrial processes, degradation of plastic objects, and the breakdown of aged microplastics (Bouwmeester et al., 2015). Microplastics comprise particles within the size range of 1–5 mm and small microplastics are those particles between 25 µm and 1 mm (Galgani et al., 2013). Microplastics can be classified as "primary" or "secondary" (see Table 2). Primary microplastics include those added to consumer products or used in industrial applications and manufacturing (e.g., microbeads in facial cleaners, toothpaste) while secondary microplastics are formed by the fragmentation of larger plastic debris (e.g., plastic bags) through environmental exposure to UV light, chemical (e.g., seawater), biological (microbial), and/or mechanical (wind and wave action) degradation (Walker, 2017; Germanov et al., 2018). The latter include synthetic fibers [the most abundant in the marine environment (Thompson et al., 2004)] in fishing gear or wastewater from textile industries and households (Germanov et al., 2018). Macroplastics are considered any plastic debris >5 mm (Moore, 2008). Macroplastics are major threats to marine life through consumption and/or entanglement (Derraik, 2002; Moore, 2008).

Some plastics such as PVC are dense enough to settle in water at the macroscopic level but PVC nanoparticles can also be found dispersed through the water column (ter Halle et al., 2017). One explanation is that nanoparticles collide with water molecules and other present ionic species that may prevent particles from sedimentation resulting in Brownian motion (Hassan et al., 2015; Gigault et al., 2018). Conversely, the

---

[1] "Any constituent that provides a molecular milieu into and onto which chemicals can escape from the aqueous solution, and whose movement is not significantly affected by gravitational settling" (Gustafsson and Gschwend, 1997).

Table 2. Sources of primary and secondary microplastics in the environment (from Arthur et al., 2009; Duis and Coors, 2016).

Microplastic type	Origin
Primary	Industrial scrubbers used for abrasive blast cleaning
	Personal care products (microbeads in cosmetics, e.g., exfoliants/abrasives)
	Medical applications (e.g., dentist tooth polish)
	Drilling fluids for oil and gas exploration
	Industrial abrasives
	Pre-production plastics, production scrap
	Accidental losses of regranulate[a], run-off from processing facilities
	Microfibers used in textiles
	Virgin resin pellets used in plastic manufacturing processes
Secondary	General plastic litter in the ocean
	Fibers from hygiene products
	Abrasion from car tyres and from synthetic textiles
	Synthetic polymers used to improve soil quality and as composting additive
	Losses of waste during waste collection, from landfill sites and recycling facilities
	Losses of plastic materials during natural disasters
	Household plastic abrasion
	Pain-related activities: paint based on synthetic polymers (ship, house, road paint, other protective paints): abrasion during use and paint removal, spills, illegal dumping
	Losses in facilities recycling plastic coated or laminated paper
	Material lost or discarded from fishing vessels and aquaculture facilities
	Material lost or discarded from merchant ships (including lost cargo), recreational boats, oil and gas platforms
	Plastic items in organic waste
	Plastic mulching[b] runoff after rain

[a] Plastic recycled from production waste.
[b] Films of low-density polyethylene (LDPE), which are used in large volumes to protect agricultural crops, suppress weed growth, increase temperature and retain irrigation water in the soil.

dispersion of macroscopic particles in seawater is governed largely by the buoyancy/sedimentation properties, directly correlated to their density, shape (e.g., film versus sphere), and adhesion/aggregation to other particles. At the microscopic scale, microplastic particles can also interact with microorganisms (bacteria, phytoplankton), which might modify their buoyancy either in a positive and negative manner (Long et al., 2015).

Plastic Degradation in the Ocean

Approximately, 46% of plastics entering the ocean are buoyant and can be dispersed over great distances by wind and ocean currents (Kubota, 1994). Although most plastics do not degrade easily in the marine environment (and on land), over time, plastics begin to break down by mechanical, photolytic, thermal, or biological degradation (Gregory and Andrady, 2003; Shah *et al.*, 2008; Worm *et al.*, 2017). Mechanical forcing through waves, inorganic salts in seawater, particles, and sediment grains can fragment plastic into smaller pieces making plastic less visible (and ultimately invisible) and altering its distribution, but not the mass of plastic in the ocean.

Like on land, where UV radiation and oxygen can lead to the formation of hydroperoxides that result in polymer chain scission over decades or centuries (Albertsson and Karlsson, 1988), exposure to UV radiation, oxygen and seawater oxidizes and breaks down plastic fragments floating on the sea surface into smaller pieces, becoming brittle (Andrady, 2005). In contrast with the oceans, plastic degradation on land occurs faster because of higher solar radiation exposure and variations in temperature (Pegram and Andrady, 1989), and plastic is often removed, for example, through active cleanups (Corcoran *et al.*, 2009).

In the ocean, plastic objects are quickly fouled by bacteria algae, reducing surface area exposure to UV radiation and oxygen soon after introduction into the marine environment (Barnes *et al.*, 2009). Microbial communities can also alter plastic buoyancy and other properties through the process of biofilm formation (Barnes *et al.*, 2009), and enzymes secreted by some microbes can cleave polymer chains of the fragment creating erosion on the plastic's surface (Shah *et al.*, 2008; Worm *et al.*, 2017). Although there are exceptions, the majority of plastic objects that reach the marine environment may remain intact for centuries, and thus accumulate in seawater in similar ways as persistent organic pollutants (POPs[2]). Although there is some evidence indicating that some microorganisms might be able to degrade secondary microplastics (e.g., after mechanical and photodegradation processes; Albertsson, 1980), and some bacteria appear to be able to degrade some types of plastic (e.g., Lu *et al.*, 2018; Tsiota *et al.*, 2018), whether microbial degradation of plastic could be used as an

---

[2]Harmful organic compounds that resist environmental degradation through chemical, biological, and photolytic processes. Rachel Carson highlights the global persistence of POPs in the environment in her book *Silent Spring* (Carson, 1962).

efficient mechanism to conduct large-scale removal of plastic from the ocean remains an open question.

Marine Plastic Pollution is a Global Problem

Marine litter is found at the sea surface and through the water column, within marine biota, coastlines, on the seafloor and in sediments, and in sea ice (Law, 2017). Both large plastics items as well as small pellets and micro- and nanoplastics can travel large distances and their distribution is irregular as it is dependent on many factors including wind and current conditions, coastline geography and points of entry into the system such as urban areas and ship trade routes. Buoyant plastics remain so until they fill with water or accumulate too much attached biota to float, thus we can see plastic litter frequently at the sea surface or washed up on the shoreline (Barnes et al., 2009). Plastic has been detected at depths below the wind-mixed layer, but in much lower concentrations than at the surface (Doyle et al., 2011).

Plastics are associated with marine life (e.g., in marine bird nests and stomachs, entangled in seals and other vertebrates, and plastic is used by hermit crabs as a substitute for shells (Barnes, 2005; Hartwig et al., 2007). However, we know very little about the extent of organisms hitchhiking on marine plastic and eventually become invaders of ecosystems (Gregory, 2009).

Although there is substantial geographic variation, plastic bags and bottles appear to account for a very high percentage (>70%) of total debris on the sea floor (Galgani et al., 2000). Beach sediments appear to harbor microplastics (small fragments, pellets, fibers) (Van Cauwenberghe et al., 2015), and deep-sea sediments contain mainly fibers (Woodall et al., 2014). Also, the rates of plastic weathering and transformation, and transport between marine reservoirs are unknown (Law, 2017), making it difficult to develop a marine plastic budget. For example, plastics that are initially buoyant may be exported to depth by biofouling and aggregation (Law, 2017). Also, some of the plastics observed in the water and on/in sediments are eventually removed via ingestion by marine animals, including seabirds (Davison and Asch, 2011; Kühn et al., 2015), intentionally removed during research or cleanup efforts, or biodegraded (Andrady, 2015; Law, 2017).

Plastics of all sizes have been found in every ocean region, converging in the subtropical gyres, including southern hemisphere gyres where, ironically, human population densities on the coast are much lower than in the northern hemisphere (Eriksen et al., 2014). Indeed, ocean gyres are considered hotspots for plastic accumulation and the buildup of plastic in these regions in on the increase. For example, a study by Goldstein et al. (2012) reported that microplastic concentrations in the North Pacific Subtropical Gyre increased by two orders of magnitude over four decades (1972–2010). Another study in 1999 by Algalita Marine Research Foundation reported that in the North Pacific Central Gyre, surface plastics outweighed zooplankton by a ratio of 6:1 and averaged over 300,000 pieces per km^2 (Moore et al., 2001). Oceanic islands are particularly vulnerable to plastic pollution as they retain plastics transported by currents and are themselves potential sources of plastics (e.g., Ivar do Sul et al., 2013). Also, coastal areas near industry are microplastic hotspots both in the water (Noren and Naustvoll, 2010) as well as in sediments (Browne et al., 2011).

Effect of Plastic on Marine Life

Plastic litter in the oceans poses a hazard for many types of organisms. For example, many marine vertebrates including fish, turtles, seabirds, and marine mammals can be easily entangled in floating plastic litter, causing reduced mobility, choking and drowning (Derraik, 2002; Kühn et al., 2015) (Table 3). Ingested plastic litter can damage internal organs, and ultimately cause death (Kühn et al., 2015). Plastic can also accumulate in many animals (bioaccumulation), for example, in filter feeders, like mussels that, when ingested by their prey, spread plastic pollution to higher trophic levels. Toxic compounds and additives of plastic can also cause major health, reproductive and hormonal imbalances in animals (Rochman et al., 2014). On the ocean floor, plastic can smother the sediment and cause hypoxia (Moore, 2008; Gregory, 2009). Moreover, plastic litter can be a vector for dispersal of organisms ultimately becoming invasive species (Lewis et al., 2005; Kiessling et al., 2015), which are considered a significant threat to coastal ecosystems (Molnar et al., 2008). Plastic can also alter organisms' behavior, for example, schools of the northern anchovy (*Engraulis mordax*) appear to be attracted to plastic debris odour causing increased aggregation and reduced rheotaxis[3] (Savoca et al., 2017).

---

[3] A common type of taxis seen in fish, whereby they will generally position themselves to face into an oncoming current.

Table 3. Compilation of peer-reviewed studies revealing impacts of marine plastic pollution (from Law, 2017).

Animal	Impact	Plastic type	Effect
Grey seals	Entanglement	Fishing line, net, rope	Compression
Elephant seals	Entanglement	Fishing line, net, rope	Dermal wound
Fur seals	Entanglement	Packing bands, fishing gear, other debris, trawl netting.	Dermal wound, death
Invertebrates	Entanglement	Derelict gillnets	Death
Fish	Entanglement	Derelict gillnets	Death
Gorgonians	Entanglement	Fishing line	Damage/breakage
Manatees	Entanglement, ingestion	Fishing line, bags, other debris	Death
Seabirds	Entanglement, ingestion	Fishing gear, fishing line, fishing hooks, derelict gillnets, plastic fragments, pellets, litter	External and internal wound, death, perforated gut
Sea turtles	Entanglement, ingestion	Fishing gear, fish hooks, plastic bags, ropes, other debris	Death, gut obstruction
Whales	Entanglement, ingestion	Plastic line, litter items, fishing gear	Dermal wound, gastric rupture, death
Penguins	Ingestion	Plastic pieces, fishing gear, other debris	Perforated gut, death
Epibenthic megafauna	Contact	Plastic bottles, glass jars	Altered assemblage
Coral reef	Contact	Lobster traps, fishing line, hook and line fishing gear	Altered assemblage, tissue abrasion
Seagrass	Contact	Crab pots, tires	Breakage, suffocation, death

Physical effects of macroplastic — entanglement and ingestion

Prior to the middle of the 20th century, fishing activities employed materials that degrade quickly (Gregory, 2009) including rope and cordage, made of natural fibers (typically Indian or Manila hemp and cotton), often strengthened with a coating of tar or strips of worn canvas. In contrast, the use of nylon and other synthetic polymers that are generally buoyant and far more durable cause major problems to marine life. An important cause of entanglement termed "ghost fishing" refers to the impact of abandoned or lost fishing gear on marine wildlife. Ghost fishing as well as plastic can cause drowning, injury, starvation, debilitation and poor quality of life in many animals. An example is plastic debris "collars" that can cause significant stress in animals, for example, in sharks and tissue abrasion as the animal grows (Sazima et al., 2002). The shape and chemistry of marine debris also govern its impact on marine life, for example, plastic packing loops such as six-pack plastic rings and plastic wires may tighten and cut into flesh as animals grow, creating "lethal necklaces" eventually leading to strangulation.

Macroplastic pieces can also be ingested by animals (Law, 2017) (Table 3), causing external and internal wounds, skin lacerations, ulcers, blockage of digestion, decrease of reproductive capacity, debilitation, drowning, hindered predator avoidance, impairment of feeding capacity, starvation and potentially death, as well as the possibility that plastic resins and additives may adsorb and be toxic (Gregory, 1991; Laist, 1997; Mato et al., 2001; Oehlmann et al., 2009; Teuten et al., 2009; Gregory, 2009). Taxa that have been identified as being affected by entanglement or ingestion of plastic, the two most reported impacts of plastic debris on wildlife, include turtles; penguins; albatrosses, petrels and shearwaters; shorebirds, skuas, gulls and auks; coastal birds other than seabirds; baleen whales, toothed whales and dolphins; earless or true seals, sea lions and fur seals; manatees and dugongs; sea otters; and fish and crustaceans (Laist, 1997; Gregory, 2009) (Table 3).

Trophic transfer and bioaccumulation of plastic

Given their inconspicuous nature, microplastics (originated from degraded macroplastics, microbeads and microplastic fibres) in the marine environment are likely a larger threat to marine systems than macroplastics (Rochman et al., 2015). Although most polymers constituting plastics have low toxicity due to their insolubility in water and because they are biochemically inert, many monomers (the constituents of polymers), such as styrene or vinyl chloride, are toxic and carcinogenic, and monomer residues in plastic products can be harmful (Lithner et al., 2011; Worm et al., 2017). For example, PVC, PU, PS, and PC, which make up ~30% of global production, contain hazardous monomers or additives (Rochman et al., 2013a) that pose health risks for humans and other species (Rochman, 2015).

Seawater harbors micro- and nanoparticles in concentrations $\sim 10^6-10^7$ particles/ml or 10–500 μg/L, most of them smaller than 100 nm in size

(Rosse and Loizeau, 2003; Andrady, 2011). A big problem is that microplastics are found in everyday products. For example, the average consumer uses facial cleansers and other products, which contain polyethylene microplastics that end up in wastewater plants and eventually enter the oceans. For example, a study on microplastics in facial cleansers available in New Zealand supermarkets found that 75% of the brands had particles <100 microns that could be ingested by plankton, and the plastic in the surface ocean subjected to UV-degradation could absorb hydrophobic materials such as polychlorinated biphenyls (PCBs), making them smaller and more toxic in the long term (Fendall and Sewell, 2009).

Due to their small size and worldwide distribution (Van Sebille *et al.*, 2015), microplastics are available to organisms at the base of the trophic chain. Many of these organisms exert limited selectivity between particles and capture anything within the size range of their prey, including plastic particles (Moore, 2008). Additionally, higher trophic planktivores passively ingest microplastics or mistake particles for natural prey. For example, microplastics found in seal feces appear to have been first accumulated in myctophid fish, which feed on copepods of the same size as the ingested plastic particles (Eriksson and Burton, 2003). Also, abundant microplastics have been found in the gastrointestinal tracts of large pinnipeds and cetaceans, indicating trophic transfer from their prey (Eriksson and Burton, 2003; Lusher *et al.*, 2015).

Microplastics can also be vectors for the transfer of toxins and POPs, which can be passed on to organisms, for example, birds, via ingestion (Worm *et al.*, 2017; Rochman *et al.*, 2014; Germanov *et al.*, 2018), e.g., when they are ingested by small planktonic organisms such as copepods and fish larvae (Carpenter *et al.*, 1972; Moore *et al.*, 2001) that are food to many predators at higher trophic levels. Indeed, small filter feeders, such as mussels, can accumulate plastic microfibers and other small particles from seawater, transferring them to benthic predators (Farrell and Nelson, 2013) and human consumers.

An example of a POP associated with plastic is the organobromine flame retardant polybrominated diphenyl ethers, which has been found in abdominal adipose tissue of marine seabirds (short-tailed shearwaters) collected in the North Pacific Ocean (Tanaka *et al.*, 2013). In long-lived filter-feeders, toxins and POPs can accumulate over decades, leading to endocrine and reproductive disruption (Rochman *et al.*, 2014). For example, plastic additives and POPs have been found in large filter feeders including the muscle of basking sharks, blubber of fin whales, and skin of whale sharks

(Fossi et al., 2012; 2014; 2017). Moreover, toxins can be transferred from mother to offspring (Lyons et al., 2013) and pose a risk to growth, survival and reproduction of the next generation.

Plastics as endocrine-disrupting chemicals

The publication of Theo Colborn's scientific bestseller *Our stolen future* (Colborn et al., 1996) represents the start of the debate of endocrine disruption as a public, political, and scientific issue. An endocrine-disrupting chemical is defined as "an exogenous substance or mixture that alters function(s) of the endocrine system and consequently produces adverse health effects in an intact organism, or its progeny, or (sub)populations" (Damstra et al., 2002). Synthetic sources of endocrine-disrupting chemicals include plastics and also detergents, drugs (e.g., oral contraceptives), cosmetics, herbicides and pesticides, as well as industrial and sewage discharges (Goksøyr, 2006).

One chemical pollutant known to disrupt the functioning of the endocrine system and associated with plastic debris is bisphenol A (BPA). This chemical has been under scrutiny because it of its estrogen-like properties and accumulation in humans as it is widely used to produce PC plastic water bottles and other resins used in food containers (Koch and Calafat, 2009). Common plasticizers such as adipates and phthalates have hormone-like properties and are frequently added to brittle plastics such as PVC to make them pliable enough for use in food packaging, toys, and many other items used daily (Koch and Calafat, 2009; Worm et al., 2017).

A number of marine studies have investigated the effect of plastic-associated chemicals on marine life. For example, a study exploring the impacts of exposure to plastic debris and associated chemicals in Japanese medaka (Oryzias latipes) revealed that the ingestion of plastic debris at environmentally relevant concentrations may alter endocrine system function in adults (Rochman et al., 2013b). In this study, the expression of genes involved in reproduction was altered in both males and females after exposure to intact polyethylene pellets and polyethylene pellets "aged" for three months in seawater. Choriogenin[4] (Chg H) gene expression was down-regulated in males and the expression of vitellogenin (Vtg I),

---

[4]Choriogenins (Chgs) are precursors of the inner layer of egg envelope that are synthesized in fish liver in response to estrogens. Chgs are sensitive biomarkers of

Chg H and the estrogen receptor (ERα) were significant down-regulated in females (Zhang et al., 2008). Similarly, exposure of marine medaka (*Oryzias melastigma*) from hatching to adulthood to the plastic softener di-(2-ethylhexyl)-phthalate (DEHP) and its derivative mono-(2-ethylhexyl)-phthalate (MEHP) caused reproductive dysfunction and endocrine disruption. The study concluded that both DEHP and MEHP cause reproductive dysfunction in marine medaka (Ye et al., 2014).

Increased incidence of disease: Effect of plastic pollution on corals

Plastic can serve as a vector of infectious diseases as shown by the presence of the pathogen *Vibrio parahaemolyticus* on a number of microplastic particles including polyethylene, polypropylene and polystyrene from North/Baltic Sea, which can harm corals (Kirstein et al., 2016). Indeed, *Vibrio coralliilyticus* sp. is a coral-pathogenic species, and strains of *V. coralliilyticus* have been isolated from diseased oyster and bivalve larvae (Ben-Haim et al., 2003).

The stress that plastic poses on coral reefs adds to their particular vulnerability to climate change, for example, warming, which causes coral bleaching (Hoegh-Guldberg, 1999; see Chapter 3); ocean acidification, that can reduce coral calcification (Jokiel et al., 2008) as well as alter microbial associations and productivity in corals (Webster et al., 2013; Zhou et al., 2016); and coastal eutrophication, which exacerbates both coral bleaching and coral disease (Vega Thurber et al., 2014). Indeed, outbreak of destructive infectious diseases and new syndromes, especially in Caribbean reefs and also in reef ecosystems in the Philippines, Australia, and East Africa have been observed over the last 20 years, as well as >50% coral loss in the Yucatán Peninsula, including pristine areas (Harvell et al., 2007). A study using 124,000 reef-building corals from 159 reefs in the Asia-Pacific region revealed that the likelihood of disease increases from 4% to 89% when corals were in contact with plastic, and that structurally complex corals are eight times more likely to be affected by plastic (Lamb et al., 2018). This study also estimated that 11.1 billion plastic items

---

exposure to estrogenic pollutants or endocrine-disrupting chemicals (Chen et al., 2008; Hong et al., 2009).

are entangled on coral reefs across the Asia-Pacific and project this number to increase by 40% by 2025.

Effect of plastic pollution on benthic systems

On top of the consequences of entanglement and ingestion, plastic litter on the benthos can impact benthic organisms by reduction of light irradiance and gas exchange between the water and sediment causing hypoxia or anoxia, thus disrupting ecosystem functioning (Goldberg, 1994; Hess et al., 1999). These conditions are known to promote the formation of polymicrobial mats characteristic of black band disease[5] (Glas et al., 2012; Lamb et al., 2018). This can cause a decline in primary productivity (because plastic causes light attenuation) and affect the amount of organic matter (Green et al., 2015) leading to changes in the seafloor structure (Gilardi et al., 2010). However, research on this type of impact of marine plastic pollution still remains in its infancy.

Effect of plastic on the expansion of invasive species: Plastisphere and plastic rafting

The huge amounts of plastic debris floating in the world's oceans remain buoyant for long periods of time and thus they are vehicles for organisms to travel long distances (e.g., Aliani and Molcard, 2003). Indeed, floating plastic produced by humans can serve as a substrate for rafting organisms ranging from microbes to sessile and mobile invertebrates, and it is also known to attract swimming animals (e.g., fish, marine vertebrates and invertebrates, dolphins, turtles) that aggregate below the plastic objects (see Kiessling et al., 2015)).

Rafting appears to be an important process for the population dynamics of many marine organisms and their dispersal over long distances (>1,000 km) (Jokiel, 1984; Castilla and Guinez, 2000; Barnes and Fraser, 2003). The hard surface of plastic provides an ideal environment for

---

[5] A disease caused by a microbial assemblage that forms a band which moves across healthy coral colonies, destroying coral tissue and exposing the bare coral skeleton. This disease causes the production of toxic sulfide, anoxia, and low pH at the boundary of the bacterial mat and the coral tissue.

opportunistic microbes to form biofilms, which might develop a niche capability for supporting diverse microorganisms, known as the "Plastisphere" (Keswani et al., 2016).

The characteristics of the polymer substrate (complexity, surface, size) are important in defining the composition of the rafting community. Many marine organisms, e.g., invertebrates, have life histories that include at least one dispersal stage, and there is evidence of benthic invertebrates living on marine debris transported by wind and surface currents (Aliani and Molcard, 2003). While rafting, organisms may be predated, compete for resources and undergo species succession. The ability to drift outside the range of the organism's ecosystem depends on their ability to hold on to floating plastic, establish themselves and compete successfully, and develop persistent local populations during their voyage (Thiel and Gutow, 2005).

A total of 1205 rafting species were reported by Thiel and Gutow (2005) including biofouling microorganisms (cyanobacteria, phytoplankton), protists, invertebrates, hydrozoans, bryozoans, crustaceans and gastropods. Organisms dispersed via plastic rafting can result in the introduction of nonnative species into new habitats (Gregory, 2009; Barnes, 2002). Therefore, this hitchhiking mechanism is of particular concern because these drifting biotopes could act as vectors for the dispersal of harmful microorganisms including pathogens, and harmful algal bloom species (Keswani et al., 2016).

Regulation of Plastic Disposal

Plastic pollution in the ocean is problematic because plastic cannot be easily removed (particularly nano and microplastic), it accumulates in organisms and sediments, and persists much longer than on land. On a local scale, although beach cleanups have some educational value, this practice is considered shortsighted, and a temporary cure at best (Williams and Tudor, 2001; Gregory, 2009). In fact, mechanical cleaning of beaches can actually cause damage, for example, via tractor-pulled sieving machines that shovel up the upper sediment. Although the near-surface meiofauna can recover quickly, repeated cleanings or deeper excavations may result in much slower recolonization rates (Gheskiere et al., 2006). Therefore, given the potential damage on species abundance and diversity (Belpaeme et al., 2004), beach cleaning is generally considered short term and the shoreline returns quickly to steady state re: plastic litter.

At the international level, attention has been focused on coordinating action to reduce large, visible (macroplastic) and small (microplastic) pollution in the marine environment (Borrelle et al., 2017). Interestingly, while the regulation of waste disposal into the ocean is very strict from marine vessels, the situation is rather different regarding the regulation of waste disposal from land (Haward, 2018). The International Maritime Organization, responsible for the administration of the International Convention for the Prevention of Pollution from Ships 1973/78 (MARPOL), Contracting Parties to the Convention on the Prevention of Marine Pollution by Dumping of Wastes and Other Matter 1972 and its 1996 Protocol (London Convention/Protocol), have recognized the problem of plastic pollution and marine litter (Haward, 2018). Accordingly, the disposal of plastics from ships is prohibited under the MARPOL convention for vessels flagged to parties to this convention in both exclusive economic zones and waters beyond national jurisdiction. Although MARPOL was signed in 1973, a complete ban on the disposal of plastics at sea was not enacted until the end of 1988. Ironically, despite 134 nations agreeing to eliminate plastic disposal at sea, oceanic sampling indicates that plastic pollution has been maintained or even worsened since MARPOL was signed (Rochman et al., 2013a) highlighting that waste disposal from land sources is in urgent need of similar monitoring and control by states as it is for waste pollution from ships.

The human population is projected to grow to nearly 10 billion by 2050 and exceed 11 billion by 2100 (United Nations, 2017), and plastic pollution is expected to continue to increase. "Given projected growth in consumption, in a business-as-usual scenario, by 2050, oceans are expected to contain more plastics than fish (by weight), and the entire plastics industry will consume 20% of total oil production, and 15% of the annual carbon budget." (https://newplasticseconomy.org/publications/report-2016). Although there have been some improvements in stabilization of plastic production and improvement of waste management and recycling practices in some jurisdictions, (1) local, regional, and international governance has failed to curtail the amount of plastic entering the marine environment, and (2) both global and local governance responses are needed to effectively manage the marine plastic litter (Vince and Hardesty, 2016). Coordinating government and non-state efforts in a holistic, integrated way has been proposed as a way forward to mitigating marine plastic pollution (Vince and Hardesty, 2016).

Today, many plastic products cannot be reused or recycled and as much as 30% of packaging products are designed for landfill or incineration (ten Brink et al., 2018). Given that the main source of plastic originates from landfills, it would make sense for producers and designers of plastic products to move away from those destined for landfills (ten Brink et al., 2018). Just in Europe, in 2006, $\sim$24.6 Mt of plastics were discarded of which more than 52% ended up in landfills, 29% were incinerated (recovering energy), and 19% were recycled (recovering materials). Eight years later, the total amount of plastic wasted did not change but less than one-third went to landfills (31%) due to a rapid growth in recycling capacity (64% increase since 2006) and waste incineration for energy production (46% increase) (Plastics Europe, 2016; Worm *et al.*, 2017). These changes were partly driven by effective legislation of plastics out of landfills by nine European countries (Austria, Belgium, Denmark, Germany, Luxembourg, Netherlands, Norway, Sweden, Switzerland) that now treat all plastic waste as a resource. However, the level of plastic consumption is unsustainable because plastics are also a source of fossil fuels and contain harmful compounds. A big part of the solution must be to vastly reduce the amount of plastic production and consumption and identify non-toxic plastic alternatives to plastic.

Some Solutions to Plastic Pollution

A big problem with plastic is that it has reached each corner of everyday life because plastic is cheap, versatile, easy to transport and light. According to The Ellen McArthur Foundation, only 14% of the plastic packaging used globally is recycled, 40% ends up in landfills and a third enters vulnerable ecosystems. At present, although educational programs may be shifting consumer's preferences to more durable products that contain little or no plastic, to expect that the consumer will cut down on plastic to make a significant difference to global plastic pollution is perhaps unrealistic.

The cost of marine litter is a big economic issue. In the Asia-Pacific region, marine litter has been estimated to be $\sim$€1 billion per year to marine industries, equivalent to 0.3% of the gross domestic product for the marine sector (McIlgorm *et al.*, 2011). As plastic production is expected to triple by 2025 (Government Office for Science, 2018), some proposals are gaining some attention:

- Implementation of extended producer responsibility (EPR) as a new standard in waste management. The Organization for Economic Co-operation and Development defines EPR as an environmental policy approach in which a producer's responsibility for a product is extended to the post consumer stage of the product's life cycle, including its final disposal (OECD, 2001).
- Removal of plastic items from the market when plastic can easily be substituted by a better product (e.g., biodegradable, non-toxic), such as

the prohibition of single-use plastics, e.g., plastic bags. For example, in 2007, San Francisco became the first U.S. city to ban single-use plastic bags, followed by more than 150 cities and counties in the U.S. that have banned bags, with nearly 50% of the US population under either a bag ban or bag charge between 2007 and 2015 (Larsen, 2017).

- Implementation of policies such as the container-deposit legislation, a deposit-refund system on drink containers (refillable or non-refillable) at the point of sale. When the container is returned to an authorized redemption center, or retailer in some jurisdictions, the deposit is partly or fully refunded to the redeemer (presumed to be the original purchaser). This not only promotes recycling but also reduces container litter, can extend the lifetime of the product, and could curb the amount of debris entering the ocean (Schuyler *et al.*, 2018).
- Implementation of policies by small and large companies and corporations with the aim of reducing plastic pollution by altering plastic production and processing. For example, Unilever has promised that by 2025, all its plastic packaging will be fully reusable, recyclable, or compostable in a commercially viable manner (https://www.unilever.com/news/press-releases/2017/Unilever-commits-to-100-percent-recyclable-plastic.html).
- Exploration of the concept of Circular Economy[6] to tackle how society can deal with increasing resource scarcity and depletion of non-renewable resources. The New Plastics Economy, an initiative led by the Ellen MacArthur Foundation, proposes a system in which all plastic materials are reused, recycled, or safely composted in a controlled way. However, given that up to a third of all plastic packaging items are too small (e.g., straws) or too complex (e.g., multimaterial films, take-away coffee cups) to be recycled in an economically viable way, achieving these goals will require a great degree of innovation, commitment and ultimately a reduction in plastic production.
- Classification of most harmful plastics as hazardous thus giving environmental agencies the power to restore affected habitats and prevent further accumulation of harmful plastic debris (Rochman *et al.*, 2013a).

---

[6] An alternative to linear economy (manufacture, consume, dispose) providing resources that can remain in use for as long as possible, obtain their maximum value and recover and regenerate products and materials at the end of their life.

- Classification of plastics in a way similar to how food is classified. For example, just as we label food as organic and free range, provide plastic consumers with explicit information about the type of plastic and aditives the consumer is buying rather than what the product does not contain (e.g., free of BPA, PVC, phthalates). Just like in the food industry, giving the consumer a choice can shift the market to less hazardous types of plastic or even a substitution of plastic by another product.

Key Points
- Plastics are the most abundant component of marine litter sometimes accounting for 100% of floating waste.
- Thermoplastics are an important source of non-biodegradable plastics to the environment accounting for ∼85% of total plastic demand in almost all market sectors.
- Marine plastic pollution is expected to triple by 2025. This rapid production of plastics and the slow growth of plastic recycling rates means that marine plastic pollution will continue to harm marine ecosystems.
- The conspicuous impacts of plastic include entanglement of marine animals and plants in plastic debris and plastic ingestion, causing wounds, starvation and ultimately death. However, nanoplastics and microplastics, originated from degraded macroplastics, microbeads and microplastic fibres, are likely a larger threat to marine life than macroplastics.
- The "Great Pacific garbage patch" and plastic pollution in coastal regions have received much attention, but the distribution and long-term impacts of the invisible plastic remain open questions.
- The invisible effects of plastic pollution are caused by compounds that can leach into the marine environment causing harm to marine organisms. Plastics such as PVC, PU, PS, and PC contain hazardous monomers, additives, and organic molecules that can bind to plastic, such as lead, DDT, and PCBs, are harmful to humans and are linked to cancer.
- Endocrine disruptors (e.g., adipates, phthalates; BPA, which are used to produce PC plastic water bottles) have hormone-like properties and are found to have harmful effects on reproduction and the endocrine system of marine animals and their consumers, including humans.
- Weathering of plastic does not make it disappear but makes it eventually invisible as microplastic or nanoplastic.

- A major source of marine plastic pollution at sea originates from land, which lacks monitoring and control by many states.

Questions

1. Why is plastic toxic to marine life?
2. Give examples of how plastic pollution can result in hypoxia and decreased light irradiance.
3. What are the main processes leading to weathering of plastic? Does plastic weathering eliminate plastic pollution?
4. Name some effects of biofouling on plastic.
5. What marine ecosystems have the highest accumulation of plastic making these biomes particularly vulnerable to plastic pollution?
6. Give an example of bioaccumulation of plastic in marine systems.
7. Explain why plastic can be a vehicle for endocrine-disrupting chemicals and give examples of these chemicals.
8. Give examples of daily routines that generate plastic pollution directly into aquatic ecosystems.
9. Give examples of how endocrine-disrupting chemicals associated with plastic can be harmful to marine organisms.
10. Name a consequence of plastic hitchhiking on marine ecosystems.

References

Alauddin M, Choudkury IA, Baradie MA, Hashmi MSJ (1995) Plastics and their machining: A review. *Materials Processing Technology* **54**, 40–46.

Albertsson A (1980) The shape of the biodegradation curve for low and high density polyethenes in prolonged series of experiments. *European Polymer Journal* **16**, 623–630.

Albertsson AC, Karlsson S (1988) The three stages in degradation of polymers — polyethylene as a model substance. *Journal of Applied Polymer Science* **35**, 1289–12302.

Aliani S, Molcard A (2003) Hitch-hiking on floating marine debris: Macrobenthic species in the Western Mediterranean Sea. *Hydrobiologia* **503**, 59–67.

American Plastics Council Year End Statistics for 2004 (http://www.americanplasticscouncil.org/sapc/docs/1700/1678.pdf).

Andrady AL (Ed.) (2005) Plastics in marine environment. A technical perspective. *Proceedings of the Plastic Rivers to Sea Conference*, Algalita Marine Research Foundation, Long Beach, California.

Andrady AL (2011) Microplastics in the marine environment. *Marine Pollution Bulletin* **62**, 1596–1605.

Andrady AL (2015) *Plastics and Environmental Sustainability.* Hoboken, NJ, John Wiley & Sons, Inc., p. 324.

Arthur C, Baker J, Bamford H (eds). (2009) *Proceedings of the International Research Workshop on the Occurrence, Effects and Fate of Microplastic Marine Debris.* September 9–11, 2008. NOAA Technical Memorandum NOS-OR&R-30.

Barnes DKA (2002) Human rubbish assists alien invasions. *Directions in Science* **1**, 107–112.

Barnes DKA (2005) Remote islands reveal rapid rise of Southern Hemisphere sea debris. *Directions in Science* **5**, 915–921.

Barnes DKA, Fraser KPP (2003) Rafting by five phyla on man-made flotsam in the Southern Ocean. *Marine Ecology Progress Series* **262**, 289–291.

Barnes DKA, Galgani F, Thompson RC, Barlaz M (2009) Accumulation and fragmentation of plastic debris in global environments. *Philosophical Transactions of the Royal Society B* **364**, 1985–1998.

Belpaeme K, Kerckhof F, Gheskiere, T (2004) About "clean" beaches and beach cleaning in Belgium. In G.R. Green (ed.), Littoral 2004: 7th *International Symposium: Delivering Sustainable Coasts: Connecting Science and Policy.* Aberdeen, Scotland, UK, 20th–22nd September 2004, Proceedings, Vol. 2, pp. 749–751.

Ben-Haim Y, Thompson FL, Thompson CC, Cnockaert MC, Hoste B, Swings J, Rosenberg E (2003) *Vibrio coralliilyticus* sp. nov., a temperature dependent pathogen of the coral *Pocillopora damicornis*. *International Journal of Systematic and Evolutionary Microbiology* **53**, 309–315.

Borrelle SB, Rochman CM, Liboiron M, Bond AL, Lusher A, Bradshaw H, Provencher JF (2017) Opinion: Why we need an international agreement on marine plastic pollution. *Proceedings of National Academic Sciences USA* **114**, 9994–9997.

Boucher C, Morin M, Bendell LI (2016) The influence of cosmetic microbeads on the sorptive behavior of cadmium and lead within intertidal sediments: A laboratory study. *Regional Studies in Marine Science* **3**, 1–7.

Bouwmeester H, Hollman PCH, Peters RJB (2015) Potential health impact of environmentally released micro- and nanoplastics in the human food production chain: Experiences from nanotoxicology. *Environmental Science and Technology* **49**, 8932e8947.

Browne MA, Crump P, Nivens SJ, Teuten E, Tonkin A, Galloway T, Thompson R (2011) Accumulation of microplastics on shorelines worldwide: Sources and sinks. *Environmental Science and Technology* **45**, 9175–9179.

Carson R (1962) Silent Spring. Houghton Mifflin Company, Boston, MA, U.S.A., 378 pp.

Carpenter EJ, Anderson SJ, Harvey GR, Miklas HP, Peck BB (1972) Polystyrene spherules in coastal waters. *Science* **178**, 749–750.

Castilla JC, Guinez R (2000) Disjoint geographical distribution of intertidal and nearshore benthic invertebrates in the southern hemisphere. *Revista Chilena De Historia Natural* **73**, 585–603.

Chen X, Li VWT, Yu RMK, Cheng SH (2008) Choriogenin mRNA as a sensitive molecular biomarker for estrogenic chemicals in developing brackish medaka (Oryzias melastigma). *Ecotoxicology and Environmental Safety* **71**, 200–208.

Colborn T, Dumanoski D, Myers JP (1996) *Our Stolen Future*, Penguin Books USA Inc, New York, p. 306.

Colton JB, Knapp FD (1974) Plastic particles in surface waters of the northwestern Atlantic. *Science* **185**, 491–497.

Contracting Parties to the Convention on the Prevention of Marine Pollution by Dumping of Wastes and Other Matter (London Convention) (1972) Final Act, 1996 Protocol And Resolutions. In: *International Legal Materials*, Vol. 36, No. 1 (January 1997), pp. 1–30, Cambridge University Press, URL: http://www.jstor.org/stable/20698637.

Corcoran PL, Beisinger MC, Grifi M (2009) Plastic and beaches: A degrading relationship. *Marine Pollution Bulletin* **58**, 80–84.

Cressey D (2016) The plastic ocean. *Nature* **536**, 263–265.

Damstra T, Barlow S, Bergman A, Kavlock RJ, van der Kraak G (eds.) (2002) Global assessment of the state-of-the-science of endocrine disruptors. Geneva, World Health Organization.

Davison P, Asch RG (2011) Plastic ingestion by mesopelagic fishes in the North Pacific Subtropical Gyre. *Marine Ecology Progress Series* **432**, 173–80.

Derraik JGB (2002) The pollution of the marine environment by plastic debris: A review. *Marine Pollution Bulletin* **44**, 842–852.

Doyle MJ, Watson W, Bowlin NM, Sheavly SB (2011) Plastic particles in coastal pelagic ecosystems of the Northeast Pacific Ocean. *Marine Environmental Research* **71**, 41–52.

Duis K, Coors A (2016) Microplastics in the aquatic and terrestrial environment: Sources (with a specific focus on personal care products), fate and effects. *Environmental Science Europe*, doi: 10.1186/s12302-015-0069-y.

Eriksen M, Lebreton LCM, Carson HS, Thiel M, Moore CJ, Borerro JC, Galgani F, Ryan PG, Reisser J (2014) Plastic pollution in the world's oceans: More than 5 trillion plastic pieces weighing over 250,000 tons afloat at sea. *PLoS ONE* **9**(12), e111913. doi: 10.1371/ journal.pone.0111913.

Eriksson C, Burton H (2003) Origins and biological accumulation of small plastic particles in fur seals from Macquarie Island. *Ambio* **32**, 380–384.

Farrell P, Nelson K (2013) Trophic level transfer of microplastic: *Mytilus edulis* (L.) to *Carcinus maenas* (L.). *Environmental Pollution* **177**, 1–3.

Fendall LS, Sewell MA (2009) Contributing to marine pollution by washing your face: Microplastics in facial cleansers. *Marine Pollution Bulletin* **58**, 1225–1228.

Fossi MC, Panti C, Guerranti C, Coppola, Giannetti M, Marsili L, Minutoli R (2012) Are baleen whales exposed to the threat of microplastics? A case study of the Mediterranean fin whale (*Balaenoptera physalus*). *Marine Pollution Bulletin* **64**, 2374–2379.

Fossi MC, Coppola D, Baini M, Giannetti M, Guerranti C, Marsili L, Panti C, de Sabata E, Clo S (2014) Large filter feeding marine organisms as indicators of microplastic in the pelagic environment: The case studies of the Mediterranean basking shark (*Cetorhinus maximus*) and fin whale (*Balaenoptera physalus*). *Marine Environmental Research* **100**, 17–24.

Fossi MC, Baini M, Panti C, Galli M, Jimenez B, Munoz-Arnanz J, Marsili L, Finoia MG, Ramirez-Macias D (2017) Are whale sharks exposed to persistent organic pollutants and plastic pollution in the Gulf of California (Mexico)? First ecotoxicological investigation using skin biopsies. *Comparative Biochemistry and Physiology, Part C* **199**, 48–58.

Galgani F, Leaute JP, Moguedet P, Souplet A, Verin Y, Carpentier A, *et al.* (2000) Litter on the sea floor along European coasts. *Marine Pollution Bulletin* **40**, 516–527.

Galgani F, Hanke G, Werner S, Vrees LD (2013) Marine litter within the European marine strategy framework directive. *ICES Journal of Marine Science* **70**, 1055–1064.

Galgani F, Hanke G, Maes T (2015) Global distribution, composition and abundance of marine litter. In: Bergmann M., Gutow L, Klages M. (eds.) *Marine Anthropogenic Litter*, Springer, Cham.

Germanov ES, Marshall AD, Bejder L, Fossi MC, Loneragan NR (2018) Microplastics: No small problem for filter-feeding megafauna. *Trends in Ecology and Evolution*, DOI: 10.1016/j.tree.2018.01.005.

Gheskiere T, Madda V, Greet P, Steven D (2006) Are strandline meiofaunal assemblages affected by a once-only mechanical; beach cleaning? Experimental findings. *Marine Environmental Research* **61**, 245–264.

Gigault J, ter Halle A, Baudrimont M, Pascal P-Y, Gauffre F, Phi T-L, El Hadri H, Grassl B, Reynaud S (2018) Current opinion: What is a nanoplastic? *Environmental Pollution* **235**, 1030–1034.

Gilardi KVK, Carlson-Bremer D, June JA, Antonelis K, Broadhurst G, Cowan T (2010) Marine species mortality in derelict fishing nets in Puget sound, WA and the cost/benefits of derelict net removal. *Marine Pollution Bulletin* **60**, 376–382.

Glas MS, Sato Y, Ulstrup KE, Bourne DG (2012) Biogeochemical conditions determine virulence of black band disease in corals. *The ISME Journal* **6**, 1526–1534.

Goksøyr A (2006) Endocrine disruptors in the marine environment: Mechanisms of toxicity and their influence on reproductive processes in fish. *Journal of Toxicology and Environmental Health, Part A* **69**, 175–184.

Goldberg ED (1994) Diamonds and plastics are forever? *Marine Pollution Bulletin* **28**, p. 466.

Goldberg ED (1997) Plasticizing the seafloor: An overview. *Environmental Technology* **18**, 195–201.

Goldstein MC, Rosenberg M, Cheng L (2012) Increased oceanic microplastic debris enhances oviposition in an endemic pelagic insect. *Biology Letters* **8**, 817–820.

Government Office for Science (2018) Foresight Future of the Sea, A Report from the Government Chief Scientific Adviser. www.gov.uk/go-science.

Green DS, Boots B, Blockley DJ, Rocha C, Thompson R (2015) Impacts of discarded plastic bags on marine assemblages and ecosystem functioning. *Environmental Science and Technology* **49**, 5380–5389.

Gregory MR (1991) The hazards of persistent marine pollution: Drift plastics and conservation islands. *Journal of the Royal Society New Zealand* **21**, 83–100.

Gregory MR (2009) Environmental implications of plastic debris in marine settings — entanglement, ingestion, smothering, hangers-on, hitch-hiking and alien invasions. *Philosophical Transactions of the Royal Society B* **364**, 2013–2025.

Gregory MR, Ryan PG (1997) Pelagic plastics and other seaborne persistent synthetic debris: A review of Southern Hemisphere perspectives. In: Coe, JM, Rogers, DB (eds.), *Marine Debris — Sources, Impacts and Solutions*. Springer-Verlag, New York, pp. 49–66.

Gregory MR, Andrady AL (2003) Plastics in the marine environment. In: AL Andrady (ed.), *Plastics and the Environment*, Hoboken, John Wiley and Sons, Inc., New Jersey, pp. 379–401.

Gustafsson C, Gschwend PM (1997) Aquatic colloids: Concepts, definitions, and current challenges. *Limnology and Oceanography* **42**, 519–528.

Hartwig E, Clemens T, Heckroth M (2007) Plastic debris as nesting material in a Kittiwake-(Rissa tridactyla)-colony at the Jammerbugt, Northwest Denmark. *Marine Pollution Bulletin* **54**, 595–597.

Harvell CD, Jordan-Dahlgren E, Merkel S, Rosenberg E, Raymundo L, Smith G, Weil E, Willis B (2007) Coral disease, environmental drivers, and the balance between coral and microbial associates. *Oceanography* **20**, 172–195.

Hassan PA, Rana S, Verma G (2015) Making sense of Brownian motion: Colloid characterization by dynamic light scattering. *Langmuir* **31**, 3–12.

Haward M (2018) Plastic pollution of the world's seas and oceans as a contemporary challenge in ocean governance. *Nature Communications* doi: 10.1038/s41467-018-03104-3.

Hess NA, Ribic CA, Vining I (1999) Benthic marine debris, with an emphasis on fishery-related items, surrounding Kodiak Island, Alaska, 1994–1996. *Marine Pollution Bulletin* **38**, 885–890.

Hiatt RA, Green Brody J (2018) Environmental determinants of breast cancer. *Annual Review of Public Health* **39**, 21.1–21.21.

Hidalgo-Ruz V, Gutow L, Thompson RC, Thiel M (2012) Microplastics in the marine environment: A review of the methods used for identification and quantification. *Environmental Science and Technology* **46**, 3060–3075.

Hoegh-Guldberg O (1999) Climate change, coral bleaching and the future of the world's coral reefs. *Marine Freshwater Research* **50**, 839–866.

Holmes LA, Turner A, Thompson RC (2012) Adsorption of trace metals to plastic resin pellets in the marine environment. *Environmental Pollution* **160**, 42–48.

Hong L, Fujita T, Wada T, Amano H, Hiramatsu N, Zhang X, Todo T, Hara A (2009) Choriogenin and vitellogenin in red lip mullet (Chelon haematocheilus): Purification, characterization, and evaluation as potential biomarkers for detecting estrogenic activity. *Comparative Biochemistry and Physiology, Part C* **149**, 9–17.

International Convention for the Prevention of Pollution from Ships (MARPOL), (1973) Reprinted in International Legal Materials (ILM), Vol. 12, American Society of International Law, Washington, DC, p. 1319. Annex V Prevention of Pollution by Garbage from Ships (entered into force 31 December 1988).

Isensee K, Valdes L (2015) *Marine Litter: Microplastics. Global Sustainable Development Report Brief.* Intergovernmental Oceanographic Commission, United Nations Education, Scientific and Cultural Organization.

Ivar do Sul JA, Costa MF, Barletta M, Cysneiros FJA (2013) Pelagic microplastics around an archipelago of the Equatorial. *Atlantic Marine Pollution Bulletin*, **75**(1–2), 305–309.

Jambeck JR, Geyer R, Wilcox C, Siegler TR, Perryman M, Andrady A, Narayan R, Law KL (2015) Plastic waste inputs from land into the ocean. *Science* **347**, 768–771.

Jokiel PL (1984) Long distance dispersal of reef corals by rafting. *Coral Reefs* **3**, 113–116.

Jokiel PL, Rodgers KS, Kuffner IB, Andersson AJ, Cox EF, Mackenzie FT (2008) Ocean acidification and calcifying reef organisms: A mesocosm investigation. *Coral Reefs* **27**, 473–483.

Jones MM (1995) Fishing debris in the Australian marine environment. *Marine Pollution Bulletin* **30**, 25–33.

Kedzierski M, Le Tilly V, Cesar G, Sire O, Bruzaud S (2017) Efficient microplastics extraction from sand. A cost effective methodology based on sodium iodide recycling. *Marine Pollution Bulletin* **115**, 120–129.

Kedzierski M, D'Almeidaa M, Magueressea A, Le Granda A, Duvala H, Cesarb G, Sirea O, Bruzauda S, Le Tilly V (2018) Threat of plastic ageing in marine environment. Adsorption/desorption of micropollutants. *Marine Pollution Bulletin* **127**, 684–694.

Keswani A, Oliver D, Gutierrez T & Quilliam R (2016) Microbial hitchhikers on marine plastic debris: Human exposure risks at bathing waters and beach environments, *Marine Environmental Research* **118**, 10–19.

Kiessling T, Gutow L, Thiel M (2015) Marine litter as habitat and dispersal vector. In: M. Bergmann et al. (eds.), *Marine Anthropogenic Litter*, doi:10.1007/978-3-319-16510-3_6.

Kirstein IV, Kirmizi S, Wichels A, Garin-Fernandez A, Erler R, Loder M, Gerdts G (2016) Dangerous hitchhikers? Evidence for potentially pathogenic Vibrio spp. on microplastic particles. *Marine Environmental Research* **120**, 1–8.

Koch HM, Calafat AM (2009) Human body burdens of chemicals used in plastic manufacture. *Philosophical Transactions of the Royal Society B: Biological Sciences* **364**, 2063–2078.

Kubota M (1994) A mechanism for the accumulation of floating marine debris north of Hawaii. *Journal of Physical Oceanography* **24**, 1059–1064.

Kühn S, Bravo Rebolledo E, van Franeker JA (2015) Deleterious effects of litter on marine life. In: Bergmann M, Gutow L, Klages M (eds.), *Marine*

Anthropogenic Litter. Springer International Publishing, pp. 75–116, https://doi.org/10.1007/978-3-319-16510-3_4.

Laist DW (1997) Impacts of marine debris: Entanglement of marine life in marine debris including a comprehensive list of species with entanglement and ingestion records. In: Coe, J.M., Rogers, D.B. (eds.), *Marine Debris — Sources, Impacts and Solutions.* Springer-Verlag, New York, pp. 99–139.

Lamb JB, Willis BL, Fiorenza EA, Couch CS, Howard R, Rader DN, True JD, Kelly LA, Ahmad A, Jompa J, Harvell CD (2018) Plastic waste associated with disease on coral reefs. *Science* **359**, 460–462.

Larsen J (2017) Plastic bag bans or fees cover 49 million Americans (http://www.earth-policy.org/data_highlights/2014/highlights49).

Law KL (2017) Plastics in the marine environment. *Annual Review of Marine Science* **9**, 205–229.

Lewis PN, Riddle MJ, Smith SDA (2005) Assisted passage or passive drift: A comparison of alternative transport mechanisms for non-indigenous coastal species into the Southern Ocean. *Antarctic Science* **17**, 183–191.

Lithner D, Larsson A, Dave G (2011) Environmental and health hazard ranking and assessment of plastic polymers based on chemical composition. *Science of the Total Environment* **409**, 3309–3324.

Long M, Moriceau B, Gallinari M, Lambert C, Huvet A, Raffray J, Soudant P (2015) Interactions between microplastics and phytoplankton aggregates: Impact on their respective fates. *Marine Chemistry* **175**, 39–46.

Lu L, Wan Z, Luo T, Fu Z, Jin Y (2018) Polystyrene microplastics induce gut microbiota dysbiosis and hepatic lipid metabolism disorder in mice. *Science of the Total Environment* **631–632**, 449–458.

Lusher AL, Hernandez-Milian G, O'Brien J, Berrow S, O'Connor I, Officer R (2015) Microplastic and macroplastic ingestion by a deep diving, oceanic cetacean: The True's beaked whale *Mesoplodon mirus. Environmental Pollution* **199**, 185–191.

Lyons K, Carlisle A, Preti A, Mull C, Blasius M, O'Sullivan J, Winkler C, Lowe CG (2013) Effects of trophic ecology and habitat use on maternal transfer of contaminants in four species of young of the year lamniform sharks. *Marine Environmental Research* **90**, 27–38.

Mato Y, Isobe T, Takada H, Kahnehiro H, Ohtake C, Kaminuma O (2001) Plastic resin pellets as a transport medium for toxic chemicals in the marine environment. *Environmental Science and Technology* **35**, 318–324.

McIlgorm, A, Campbell, HF, Rule, MJ (2011) The economic cost and control of marine debris damage in the Asia-Pacific region. *Ocean and Coastal Management* **54**, 643–651.

Molnar JL, Gamboa RL, Revenga C, Spalding, MD (2008) Assessing the global threat of invasive species to marine biodiversity. *Frontiers in Ecology and the Environment* **6**, 485–492.

Moore CJ (2008) Synthetic polymers in the marine environment: A rapidly increasing, long-term threat. *Environmental Research* **108**, 131–139.

Moore CJ, Moore SL, Leecaster MK, Weisberg SB (2001) A comparison of plastic and plankton in the North Pacific Central Gyre. *Marine Pollution Bulletin* **42**, 1297–1300.

Muller RJ, Kleeberg I, Deckwer WD (2001) Biodegradation of polyesters containing aromatic constituents. *Journal of Biotechnology* **86**, 87–95.

Nash A (1992) Impacts of marine debris on subsistence fishermen — an exploratory study. *Marine Pollution Bulletin* **24**, 150–156.

Noren F, Naustvoll F (2010) Survey of microscopic anthropogenic particles in Skagerrak. Commissioned by KLIMA-OG FORURENSNINGSDIREKTORATET, Norway.

OECD (2001) Extended producer responsibility: A guidance manual for governments. Paris7 OECD.

Oehlmann J, Schulte-Oehlmann U, Kloas W, Jagnytsch O, Lutz I, Kusk KO, Wollenberger L, Santos EM, Paull GC, Van Look KJW, Tyler CR (2009) A critical analysis of the biological impacts of plasticizers on wildlife. *Philosophical Transactions of Royal Society B* **364**, 2047–2062.

Pegram JE, Andrady AL (1989) Outdoor weathering of selected polymeric materials under marine exposure conditions. *Polymer Degradation and Stability*, **26**, 333–345.

Plastics Europe (2015) Plastics — the facts. An analysis of European plastics production, demand and waste data. Brussels, Belgium.

Plastics Europe (2016) Plastics — the facts. An analysis of European plastics production, demand and waste data. Brussels, Belgium, Plastics Europe, http://www.plasticseurope.org/documents/document/20161014113313-plastics_the_facts_2016_final_version.pdf.

Ramirez-Llodra E, De Mol B, Company JB, Coll M, Sarda F (2013) Effects of natural and anthropogenic processes in the distribution of marine litter in the deep Mediterranean Sea. *Progress in Oceanography* **118**, 273–287.

Rios LM, Moore C, Jones PR (2007) Persistent organic pollutants carried by synthetic polymers in the ocean environment. *Marine Pollution Bulletin* **54**, 1230–1237.

Rochman CM (2015) The complex mixture, fate and toxicity of chemicals associated with plastic debris in the marine environment. See Ref. **109**, 117–40.

Rochman CM, Browne MA, Halpern BS, Hentschel BT, Hoh E, Karapanagioti HK, Rios Mendoza LM, Takada H, Teh SJ, Thompson RC (2013a) Classify plastic waste as hazardous. *Nature* **494**, 169–171.

Rochman CM, Hoh E, Kurobe T, Teh SJ (2013b) Ingested plastic transfers hazardous chemicals to fish and induces hepatic stress. *Nature Scientific Reports*, doi: 10.1038/srep03263.

Rochman CM, Kurobe T, Flores I, Teh SJ (2014) Early warning signs of endocrine disruption in adult fish from the ingestion of polyethylene with and without sorbed chemical pollutants from the marine environment. *Science of the Total Environment* **493**, 656–661.

Rochman CM, Tahir A, Williams SL, Baxa DV, Lam R, Miller JT, Teh FC, Werorilangi S, Teh, Werorilangi S, Teh SJ (2015) Anthropogenic debris in seafood: Plastic debris and fibers from textiles in fish and bivalves sold for human consumption. *Scientific Reports* **5**, 14340. doi: 10.1038/srep14340.

Rosse P, Loizeau J-L (2003) Use of single particle counters for the determination of the number and size distribution of colloids in natural surface waters. *Colloids Surface A* **217**, 109–120.

Savoca MS, Tyson CW, McGill M, Slager CJ (2017) Odours from marine plastic debris induce food search behaviours in a forage fish. *Proceedings of Royal Society B* **284**, 20171000. http://dx.doi.org/10.1098/rspb.2017.1000

Sazima I, Gadig OBF, Namora RC, Motta FS (2002) Plastic debris collars on juvenile carcharhinid sharks (*Rhizoprionodon lalandii*) in southwest Atlantic. *Marine Pollution Bulletin* **44**, 1147–1149.

Schmidt C, Krauth T, Wagner S (2017) Export of plastic debris by rivers into the sea. *Environmental Science and Technology* **51**, 12246–12253.

Schuyler Q, Hardesty BD, Lawson TJ, Opie K, Wilcox C (2018) Economic incentives reduce plastic inputs to the ocean. *Marine Policy*, doi.org/10.1016/j.marpol.2018.02.009.

Shah AA, Hasan F, Hameed A, Ahmed S (2008) Biological degradation of plastic: A comprehensive review. *Biotechnology Advances* **26**, 246–265.

SPI (Society of Plastics Industry) (2015) History of plastics. https://www.plasticsindustry.org/AboutPlastics/content.cfm?ItemNumber=670&navItemNumber=1117.

Tanaka K, Takada H, Yamashita R, Mizukawa K, Fukuwaka MA, Watanuki Y (2013) Accumulation of plastic-derived chemicals in tissues of seabirds ingesting marine plastics. *Marine Pollution Bulletin* **69**, 219–222.

ten Brink P, Schweitzer J-P, Watkins E, Janssens C, De Smet M, Leslie H, Galgani F (2018) Circular economy measures to keep plastics and their value in the economy, avoid waste and reduce marine litter. Economics Discussion Papers, No. 2018–3.

ter Halle A, Ladirat L, Martignac M, Mingotaud AF, Boyron O, Perez E (2017) To what extent are microplastics from the open ocean weathered? *Environmental Pollution* **227**, 167–174.

Teuten EL, Saquing JM, Knappe DRU, Barlaz MA, Jonsson S, Björn A, Rowland SJ, Thompson RC, Galloway TS, Yamashita R, Ochi D, Watanuki Y, Moore C, Viet PH, Tana TS, Prudente M, Boonyatumanond R, Zakaria MP, Akkhavong K, Ogata Y, Hirai H, Iwasa S, Mizukawa K, Hagino Y, Imamura A, Saha M, Takada H (2009) Transport and release of chemicals from plastics to the environment and to wildlife. *Philosophical Transactions of Royal Society B* **364**, 2027–2045.

The American Chemistry Council (2015) 2015 Statistical Reference Book on Plastic Resins, *The Resin Review*, Washington, DC.

Thiel M, Gutow L (2005) The ecology of rafting in the marine environment. II. The rafting organisms and community. *Oceanography and Marine Biology: An Annual Review* **43**, 279–418.

Thompson RC (2017) *Future of the Sea: Plastic Pollution*. Government Office for Science, London UK.

Thompson RC, Olsen Y, Mitchell RP, Davis A, Rowland SJ, John AWG, McGonigle D, Russell AE (2004) Lost at sea: Where is all the plastic? *Science* **304**, 838.

Thompson RC, Swan SH, Moore CJ, vom Saal FS (2009) Our plastic age. *Philosophical Transactions of the Royal Society B: Biological Sciences* **364**, 1973–1976.

Tsiota P, Karkanorachaki K, Syranidou E, Franchini M, Kalogerakis N (2018) Microbial degradation of HDPE secondary microplastics: Preliminary results. In: Cocca M., Di Pace E., Errico M., Gentile G., Montarsolo A., Mossotti R. (eds.), *Proceedings of the International Conference on Microplastic Pollution in the Mediterranean Sea*. Springer Water, Springer, Cham. https://doi.org/10.1007/978-3-319-71279-6_24.

Uneputty P, Evans SM (1997) The impact of plastic debris on the biota of tidal flats in Ambon Bay (Eastern Indonesia). *Marine Environmental Research* **44**, 233–242.

United Nations Department of Economic and Social Affairs (2017) World Population Prospects: The 2017 Revision, *Key Findings and Advance Tables*. Working Paper No. ESA/P/WP/248, Department of Economic and Social Affairs, Population Division, New York.

US EPA (2006) *Municipal Solid Waste in the United States: 2005 Facts and Figures*. EPA530-R-06-011, United States Environmental Protection Agency, Office of Solid Waste, Washington, DC.

US Environment Protection Agency (2014) Municipal solid waste generation, recycling, and disposal in the United States, Tables and Figures for 2012. Rep., US EPA, Washington, DC.

Van Cauwenberghe L, Devriese L, Galgani F, Robbens J, Janssen CR (2015) Microplastics in sediments: A review of techniques, occurrence and effects. *Marine Environmental Research* **111**, 5–17.

van Sebille E, Wilcox C, Lebreton L, Maximenko N, Hardesty BD, van Franeker JA, Eriksen M, Siegel D, Galgani F, Law KL (2015) A global inventory of small floating plastic debris. *Environmental Research Letter* **10**, 124006.

Vega Thurber RL, Burkepile DE, Fuchs C, Shantz AA, McMinds R, Zaneveld JR (2014) Chronic nutrient enrichment increases prevalence and severity of coral disease and bleaching. *Global Change Biology* **20**, 544–554.

Vince J, Hardesty BD (2016) Plastic pollution challenges in marine and coastal environments: From local to global governance. *Restoration Ecology* doi:10.1111/rec.12388.

Walker TR (2017) Drowning in debris: Solutions for a global pervasive marine pollution problem. *Marine Pollution Bulleting* **126**, 338.

Webster NS, Negri AP, Flores F, Humphrey C, Soo R, Botte ES, Vogel N, Uthickle S (2013) Near-future ocean acidification causes differences in microbial associations within diverse coral reef taxa. *Environmental Microbiology Reports* **5**, 243–251.

Williams AT, Tudor DT (2001) Temporal trends in litter dynamics at a pebble pocket beach. *Journal of Coastal Research* **17**, 137–145.

Woodall LC, Sanchez-Vidal A, Canals M, Paterson GLJ, Coppock R, Sleight V, Calafat A, Rogers AD, Narayanaswamy BE, Thompson RC (2014) The deep sea is a major sink for microplastic debris. *Royal Society Open Science* **1**, 140317.

Worm B, Lotze HK, Jubinville I, Wilcox C, Jambeck J (2017) Plastic as a persistent marine pollutant. *Annual Review of Environmental Resources* **42**, 1–26.

Ye T, Kang M, Huang Q, Fang C, Chen Y, Shen H, Dong S (2014) Exposure to DEHP and MEHP from hatching to adulthood causes reproductive dysfunction and endocrine disruption in marine medaka (*Oryzias melastigma*). *Aquatic Toxicology* **146**, 115–126.

Zhang X, Hecker M, Jones PD, Newsted J, Au D, Kong R, Wu RSS, Giesy JP (2008) Responses of the *Medaka* HPG axis PCR array and reproduction to Prochloraz and Ketoconazole. *Environmental Science and Technology* **42**, 6762–6769.

Zheng, Ernest K, Yanful, Amarjeet S, Bassi (2005) A review of plastic waste biodegradation, *Critical Reviews in Biotechnology* **25**(4), 243–250.

Zhou G, Yuan T, CAi L, Zhang W, Tian R, Tong H, Jiang L, Yuan X, Liu S, Qian P, Huang H (2016) Changes in microbial communities, photosynthesis and calcification of the coral. *Acropora gemmifera* in response to ocean acidification. *Scientific Reports* **6**, 35971.

Chapter 6

Oil Pollution

The Release of Petroleum Hydrocarbon into the Ocean from Its Extraction, Transportation, Refining, Storage, and Use Is Harming Marine Life

"By recklessly expanding offshore oil drilling, we risk devastating impacts to our coastal economies, our state, our wildlife and our way of life. We are now seeing an unprecedented number of members of the business community joining us in speaking out in opposition to this plan. As California deepens its commitments to a clean energy future, we are tackling climate change, not denying it, and I will continue to move forward with my Senate Bill 834 to ensure that pipelines and other infrastructure cannot be built in California waters to enable the dirty, destructive and dangerous energy policies of our Federal Administration."

— Statement by Senator Hannah-Beth Jackson (D-Santa Barbara, California, USA) opposing offshore drilling along the California coast, February 8, 2018.

Introduction

Crude oil (naturally occurring product, that is used to make gasoline and diesel) is a widespread pollutant in the marine environment that comprises organic compounds including alkanes, aromatics, asphaltenes, resins and waxes, as well as various gases, solids, and trace minerals mixed in (Wang and Fingas, 2003; Marshall and Rodgers, 2004). Oil can exist in liquid, gas, and semisolid (tar or asphalt) states, and each of these states has distinct effects on marine biota, from microbial communities to populations of large animals like whales or dolphins. The composition of crude oil, its physico–chemical properties (e.g., viscosity, solubility, and capacity to absorb), bioavailability, and toxicity are highly variable (McGenity et al., 2012). The main components of crude oil are alkanes with different chain lengths and branch points, cycloalkanes, and mono-aromatic and polycyclic aromatic hydrocarbons (PAHs) (Harayama et al., 1999). Some oil compounds contain nitrogen, sulfur, and oxygen as well as trace amounts of phosphorus and heavy metals (Harayama et al., 1999; van Hamme et al., 2003).

Oceans and coastal zones are impacted by oil pollution, which affects marine chemistry and ecosystems in major ways. Every year, an estimated 1.3 million tonnes (Mt) of crude oil and petroleum (refined crude oil) enter the marine environment by natural oil seepage and by human activities, for example, via the extraction, transportation, refining, storage, and use of crude oil and natural gas (NRC, 2003). Oil spills during these operations are common in the marine environment. In addition to these frequent spills, major spills caused by failures in tankers and pipelines, which transport crude oil from one place to another, take place sporadically, most unfortunately resulting in environmental disasters. Notorious examples include the Deepwater Horizon oil rig explosion in the Gulf of Mexico in 2010, which released 800 million liters of crude oil into the environment; the Exxon Valdez oil spill from a tanker in Alaska in 1989, which resulted in the release of 42 million liters of crude oil; and the sinking of the 26 year old structurally deficient Prestige oil tanker in Spain in 2002, which leaked 60 million liters of heavy fuel oil into the waters outside of Galicia.

Major concerns with oil pollution include the physical impacts of oil on the surface ocean, sea floor, beaches and coastal environments as well as the "invisible" carcinogenic and genotoxic effects of many oil components (De Flora et al., 1991). The biological impacts span from moderate to sub lethal and lethal and include impairment of physiological functions, such as reproduction, caused by the physical and toxic impacts of oil. This chapter discusses the state of knowledge concerning the natural or human-induced release of crude oil and refined/processed petroleum into the marine environment. The gaseous components of oil (natural gas) will not be considered in any detail here, because there are large uncertainties about rates of gas seepage and production.

What Constitutes Oil?

Crude oil contains ~17,000 compounds in liquid and volatile phases. Oil is essentially made of dead animal and plant matter, buried under deep layers of sedimentary rock that over time, and with the aid of pressure and heat, cause oil deposits to form. Specifically, organic matter is deposited on the ocean floor and buried within the sediment, where it is locked away from oxygen, thus preventing the oxidation of organic matter. Under anoxic (oxygen depleted) conditions and under high pressure and temperature, organic carbon is incorporated into sedimentary rocks and, through the mediation of catalysts, the organic detritus is slowly transformed into oil. The higher molecular weight compounds appear to be broken down over time resulting in the formation of paraffins and the production of methane (Andreev et al., 1968). Oil can remain underground for millions of years and during this time, it continues to undergo transformations.

A variety of approaches have been employed to characterize the composition of oil including gas chromatography, gas chromatography–mass spectrometry, high-performance liquid chromatography (HPLC), size-exclusion HPLC, infrared spectroscopy, supercritical fluid chromatography, thin layer chromatography, ultraviolet and fluorescence spectroscopy, isotope ratio mass spectrometry, and gravimetric methods (Wang and Fingas, 2003) (see examples in Table 1). The composition of crude oil is critically dependent on conditions during its formation and the age of the oil. Hydrocarbons (compounds made of carbon and hydrogen) make up a large part of crude oil and the most abundant and studied hydrocarbons in oil are alkanes, naphthenes, and compounds containing benzene rings. Among the components of crude oil, there is a group of molecules called PAHs, which contain 2–8 conjugated ring systems (Figure 1) and substituents including alkyl, nitro, and amino groups in their structure as well as nitrogen, sulfur, and oxygen atoms (Fieser and Fieser, 1956). PAHs have been the focus of many studies as this family of compounds is considered particularly toxic to marine life.

Natural Oil Sources: Oil Seeps

Oil seeps are sites where natural sources of crude oil leak from fractures in the seafloor and move upwards from the geologic strata beneath the seafloor to the sea surface. They harbor economic reserves of petroleum as they release vast amounts of crude oil annually and, consequently, oil (and gas) extraction activities are often concentrated in regions where seeps

Table 1. Standard procedures for determining PAHs in aqueous samples (from Wolska, 2008; Pampanin and Sydnes, 2013).

Method ID	Application	Sample preparation	Determination technique
EPA-610	PAHs in municipal and industrial wastewaters	~1L of water extracted with dichloromethane, the extract is then dried and concentrated to a volume <10 mL	HPLC or GC (with a packed column)
EPA-8310	Groundwater and wastewater	EPA 3500 series methods (version 3, 2000)	HPLC
EPA-8270D	Aqueous samples	EPA 3500 series methods (version 3, 2000)	GC-MS with use of high-resolution capillary columns and deuterated internal standard
ISO 17993:2002	Water quality (determination of 15 PAHs)	Water samples collected in brown glass bottles and stabilized by adding sodium thiosulfate. 1L of sample extracted using hexane. The extract is dried with sodium sulfate and then enriched by removal of hexane by rotary evaporation	HPLC and fluorescence detector

form. The seepage of oil takes place at low enough rates such that the surrounding ecosystem can adapt and even thrive in the presence of oil. On average, natural seeps contribute an annual release of ~47% of crude oil into the marine environment compared to ~53% resulting from petroleum leaks and spills during the extraction, transportation, refining, storage, and use of petroleum (Judd, 2003; Kvenvolden and Cooper, 2003). In North America, the largest and best-studied natural seep ecosystems are in the Gulf of Mexico and the waters off of Southern California, and indeed these regions undergo extensive oil and gas production. Although methods to calculate oil-seepage rates are far from adequate, current estimates are between 0.2

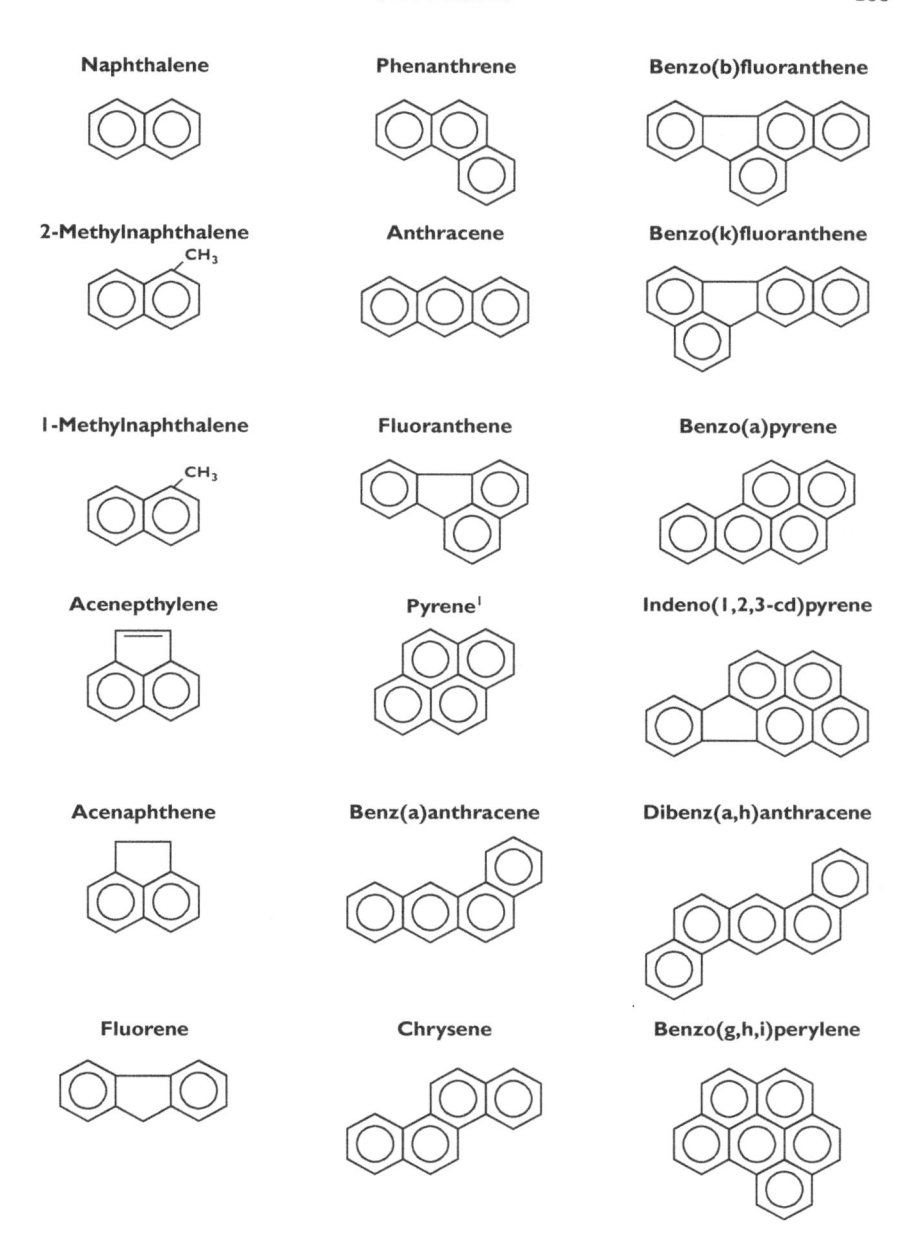

Figure 1. Structure of common polycyclic aromatic hydrocarbons.

and 2 Mt/yr, with a best estimate of 0.6 Mt/yr (Kvenvolden and Cooper, 2003).

The release of hydrocarbons through surface sediments alters benthic ecology and seafloor geology. As oil moves towards the sea surface through the water column, its chemical alteration is modest, but once it reaches the surface, volatile molecules rapidly evaporate to the atmosphere and the oil forms slicks that can be detected by remote sensing. Gas bubbles rapidly dissolve as they rise towards the surface, although observations suggest that oil coatings on the bubbles can inhibit dissolution. Ecosystems harboring natural oil seeps promote the presence of hydrocarbon-eating microorganisms (Head et al., 2006), which speed up the degradation of oil (Valentine et al., 2012).

Satellite remote sensing, synthetic aperture radar (SAR), and visible band images from Landsat TM and the space shuttle are increasingly revealing new areas with significant oil seeps. Although satellite data allow for observations of oil seeps over large spatial and temporal scales, several uncertainties make it difficult to quantify their magnitude. These include slick thickness, required to estimate total volume; residence time of oil on the surface, which is highly dependent on wind and slick patterns; and seepage rate, which cannot be considered constant as it can be episodic and ephemeral.

Pollution Generated by the Production of Petroleum and Oil-Related Activities

Extraction of oil

Oil platforms

Throughout the world's oceans, there are over 8,000 platforms and offshore facilities that both purposefully and accidentally release oil into marine waters. Globally, the estimated amount of discharge is at least 290 Mt/y (Table 2), assuming that the volume of spills/platform/year for non-US platforms is similar to those in North American waters (NRC, 2003). Despite attempts to decrease the amount of oil released during extraction activities, the potential for significant spills, especially in aged production fields with old infrastructures, is still a threat to the environment.

Produced water

A byproduct of oil and gas production offshore and onshore is large volumes of water containing small amounts of oil and other added chemicals called

Table 2. Estimates of average releases of oil sources between 1990 and 1999 in thousands of tonnes per year from NRC (2003).

Source of oil	Best estimate	Estimated range
Natural seeps	600	200–2,000
Extraction of petroleum	38	20–62
Platforms	0.86	0.29–1.4
Atmospheric deposition	1.3	0.38–2.6
Produced waters	36	19–58
Transportation of petroleum	150	120–260
Pipeline spills	12	6.1–37
Tanker spills	100	93–130
Operational discharges (cargo washing)	36	18–72
Coastal facility spills	4.9	2.4–15
Atmospheric deposition	0.4	0.2–1
Consumption of petroleum	480	130–6,000
Land-based (river and runoff)	140	6.8–5,000
Recreational marine vessels	nd[1]	nd[1]
Spills (non-tankers)	7.1	6.5–8.8
Operational discharges (vessels ≥100 GT)	270	90–810
Operational discharges (vessels <100 GT)	nd[2]	nd[2]
Atmospheric deposition	52	23–200
Jettisoned aircraft fuel	7.5	5.0–22
Total	1300	470–8,300

produced water (PW). PW generally consists of a mixture of (1) water contained naturally in the reservoir; (2) injected water used for the recovery of oil; and (3) treatment chemicals added during oil extraction (Røe Utvik, 1999). There is growing concern about PW and its possible long-term impact on the environment as PW is released on a regular basis from offshore oil installations and the volume of PW discharges can be significant (Furnes, 1994). For example, although the amount of oil in PW is regulated, even low concentrations of PAHs contained in PW are likely to cause adverse effects in the marine environment (Sundt *et al.*, 2012). Moreover, at shore-based facilities, the discharge of treated PW derived from many offshore platforms is still permitted in some U.S. locations, for example, Upper Cook Inlet, Alaska, while the discharge of PW into estuaries and shallow coastal waters continues globally in developing fields (e.g., Nigeria, Angola, China, Thailand) (NRC, 2003).

Atmospheric deposition

During the production, transport, and refining of crude oil, volatile compounds are released to the atmosphere, a proportion of which are termed volatile organic compounds (VOCs) and are defined by the U.S. Clean Air Act as a family of compounds that include all volatile hydrocarbons except methane, ethane, a wide range of chlorofluorocarbons, and hydrochlorofluorocarbons. These compounds are of concern due to their impact on air quality and human health, having the potential to cause eye, nose and throat irritation, headaches, nausea, damage to liver, kidney and central nervous system, and cancer (Solomon and Janssen, 2010). Methane is light and mixes or degrades rapidly while high molecular weight compounds, such as hexadecane, react slowly and may be deposited on the sea surface.

Transportation of petroleum

Pipeline spills

Pipelines transport oil from offshore production sites to coastal refineries and distribution installations. In some parts of the world, pipelines are the main method of oil transport, for example, almost 60% of the oil produced in the North Sea is delivered to land via pipeline networks. Recently, a sharp increase in accidental discharges from pipelines appears to have been caused by three main reasons — the increase in the number and total length of oil pipelines since the 1970s and hence enhanced probability of accidents, the ageing of pipelines and pumping stations, and the inadequate maintenance, and corrosive conditions that can lead to pipeline ruptures (Figure 2). Additionally, a number of pipelines have become military targets in uprisings and tribal wars, such as in the delta of the Niger River and parts of the Amazon (Jernelöv, 2010). Operational discharges of oil from

Figure 2. Images from the 2015 Refugio oil spill (Santa Barbara, CA, 2015). An underground pipeline near Refugio State Beach (∼32 km west of Santa Barbara, CA, USA) leaked between 101,000 and 140,000 gallons (∼382 327 to 529 958 l) of oil of which ∼21,000 gallons (79,494 L) entered the ocean (www.refugioresponse.com/go/doc/7258/2588430/). Images were kindly provided by Dr. Anna James.

pipelines are usually relatively small and often occur during commissioning, maintenance and decommissioning phases although there are no published data on the volume or the rate of discharges (GESAMP, 2007).

Oil tanker spills

In contrast with spills associated with oil extraction, which tend to be concentrated near production fields, oil (including VOCs) spills associated with transportation can occur anywhere tankers travel or where pipelines are located. Also, spills from transportation activities may release a wide variety of petroleum products (not just crude oil) each of which has a different fate in the environment. This is because some distillates evaporate faster than others and different oil types contain different concentrations of toxic compounds like PAHs.

Operational discharges

There used to be two major sources of operational discharges: oil-contaminated ballast water and cargo tank washings. When tankers offload cargo and prepare to travel empty, large quantities of ballast water are necessary to maintain the balance of the ship at sea. In the past, tankers stored much of this ballast water in empty cargo tanks, but because these tanks contained oil residues, the ballast water became contaminated. Therefore, the discharge of this ballast water resulted in oil pollution. Nowadays, with the exception of old tankers, segregated ballast tanks are used exclusively for ballast water (Hampton *et al.*, 2003).

Among the tankers that arrive in port bringing crude oil or refined petroleum products, they may leave with empty tanks to retrieve more of the same cargo or, after off-loading their cargo of crude oil, may reload with a refined oil product. In this latter situation, when switching cargoes, there is often a requirement to clean the cargo tanks of residue left over from the previous cargo. This is particularly important when switching from crude oil to a refined product although, even when not switching cargo or going into dry dock, regular tank washings are necessary to prevent the buildup of sludge in the tanks (Griffin, 1994). Washing each tank takes two to three days, discharging as much as 40,000 liters of oil. Typically, a tanker may have twelve or more tanks, requiring washing two to three times per month. Tank washings are conducted at sea because the vapors and fumes released during the process violate air quality standards in urban areas where ports are located. These procedures may involve different physical, chemical, and biological methods in order to remove oil residues from within the cargo tanks, which end up in seawater (Ahmadun et al., 2009). The washing of cargo tanks represents a huge indirect source of oil pollution threat to marine life, with a best average estimate of 36,000 tonnes of petroleum per year (Table 2).

Facility spills

Two main types of facilities (excluding transportation vessels) appear to be the major contributors to oil spills over a 10-year study (1990–1999): coastal pipelines transporting refined products to and from marine terminals and industrial facilities (NRC, 2003).

Environmental hazards of oil refineries

Oil refineries pose serious environmental concerns because the techniques required to separate crude oil into petroleum products (e.g., gasoline, kerosene, diesel fuel, jet fuel, waxes, asphalt) involve physical and chemical separation techniques (e.g., fractionation, cracking, hydrotreating, and manufacturing and transport) that are harmful to the environment. Some of the chemicals released during processing are known to cause cancer, developmental and reproductive problems, as well as cause or worsen certain respiratory conditions such as asthma.

The United States is the largest producer and consumer of oil in the world, and energy production in the U.S. represents ~76% of its consumption (Table 3). Oil refineries are a major source of toxic air pollutants posing great consequences to humans and the environment. Benzene, toluene, ethylbenzene, and xylene (BTEX) are an important group of VOCs because of their role in the troposphere chemistry and the health risks to humans. Although this chapter is concerned with the marine environment, it is important to note that short-term exposure to some VOCs present in the air is not considered acutely harmful to human health but also to animal welfare. On the other hand, chronic exposure may lead to mutagenic and carcinogenic effects as well as neurological disorders, fatigue, headaches, dizziness, nausea, lethargy, depression, and even developmental and reproductive problems (ATSDR, 1995; Weschler and Shields, 1997; Rumchev et al., 2004; Yang et al., 2004). Oil refineries are also a source of particulate pollutants, nitrogen oxides (NO_x), carbon monoxide (CO), hydrogen sulfide (H_2S) and sulfur dioxide (SO_2), and methane. Finally, VOCs are chemical precursors for ozone, which is formed from photochemical interactions of VOCs and oxides of nitrogen and, when

Table 3. The 10 largest producers and consumers of oil and other petroleum products and share of total world oil production (domestic production of crude oil, all other petroleum liquids) in 2016 (U.S. Energy Information Administration, 2016, https://www.eia.gov/tools/faqs/faq.php?id=709&t=6). Numbers are in millions of barrels per day.

Country	Producers	% world	Country	Consumers	% world
United States	14.85	15	USA	19.53	20
Saudi Arabia	12.39	13	China	12.02	13
Russia	11.24	12	India	4.14	4
China	4.87	5	Japan	4.12	4
Canada	4.59	5	Russia	3.55	4
Iraq	4.45	5	Saudi Arabia	3.24	3
Iran	4.21	4	Brazil	2.99	3
United Arab Emirates	3.77	4	South Korea	2.41	3
Brazil	3.24	3	Canada	2.37	3
Kuwait	3.07	3	Germany	2.37	2
Total top 10	66.68	69	—	56.78	60
World total	97.23	—	—	—	—

occurring at high levels, ozone is a major environmental pollutant which increases risks of mortality from respiratory diseases in humans (Jerrett et al., 2009; McDonald et al., 2018).

Activities in oil refineries can also contaminate ground and surface water. For example, although water is recycled through many stages during the refining process and goes through several treatment processes, some oil residues in wastewater can end up in aquifers and groundwater. Soil contamination from dust, residues in tank bottoms, and sludges from treatment processes can occur due to leaks, accidents, or spills on site or during the transport process. However, this type of contamination is generally less significant than air and water pollution from refineries.

Abiotic Transformation of Oil in the Water

Crude oil and refined products undergo physical, chemical, and biological processes that change their composition, physical and chemical properties, and environmental impact in the marine environment. Processes leading to oil transformation include evaporation, spreading, dissolution, dispersion, emulsification, photodegradation, biodegradation, sedimentation, oxidation, and the formation of tarballs that can become stranded on the shoreline. The dispersal of crude oil in seawater is governed by mechanisms that carry oil vertically and horizontally including water currents, wind driven mixing, aggregation (e.g., with other particles or with itself), water currents, sinking, and sedimentation. Weathering (e.g., evaporation, dispersion, microbial degradation, photo-oxidation) removes the lighter compounds of oil whereas the heavier components can sink to the ocean floor, where the oil residues can persist for decades (Owens et al., 2008). Also, vertical sinking after oil spills and subsequent burial in sediments change both the concentration and composition of oil in the water column, and natural processes oxidize and reduce organic compounds into CO_2 and biomass.

The chemical and physical properties of the water-soluble fraction of crude oil, dominated by the more water-soluble aromatic hydrocarbons, are different from the original released oil (Robotham and Gill, 1989). The dissolved fraction of oil rather than the emulsified or the absorbed fraction is readily adsorbed or ingested by organisms and has highly toxic effects (Robotham and Gill, 1989). Photochemical oxidation results in the generation of oxidized photoproducts that also have toxic effects on marine biota (Tilseth et al., 1984). For example, the toxicity of PAHs can increase by up to 1,000-fold in the presence of ultraviolet light (Barron and Ka'Aihue, 2001).

Biological Degradation of Oil

Oil is degraded by phylogenetically diverse microorganisms under both oxic and anoxic conditions (Head et al., 2006; Widdel et al., 2010). Bacteria are considered the main agents in environmental hydrocarbon degradation (Leahy and Colwell, 1990). There is also evidence indicating that one red alga and several diatoms can degrade hydrocarbons, although their contribution to oil degradation and environmental significance are unclear (Prince, 2010). Marine hydrocarbon-degrading bacteria are considered ubiquitous and include members of *Alpha-, Beta-* and *Gammaproteobacteria*, and Gram positives (Vila et al., 2010). Among aerobic hydrocarbon degraders in the water column are Gammaproteobacteria and Alphaproteobacteria (Kleindienst and Joye, 2017). In marine sediments, anaerobic hydrocarbon degraders mainly belong to sulfate-reducing bacteria (Widdel et al., 2010), specifically *Delta-proteobacteria* (Kleindienst and Joye, 2017). Microorganisms appear to degrade the lighter fractions of oil first, e.g., short-chain alkanes which are easily accessible energy sources (Atlas, 2011) although many microbes can also break down long-chain alkanes and PAHs (Pothuluri and Cerniglia, 1994).

Microbial degradation of oil is dependent on many factors including the oil composition, degree of weathering, water temperature, and dissolved nutrient concentrations. For example, microbial degradation may be slow in tropical and subtropical latitudes when nutrients are limited while photodegradation may be more important in these environments with elevated solar radiation and temperature (Al-Lihaibi, 2003). Photo-oxidation turns the oil components into more water-soluble oxygenated molecules that is likely to accelerate the dispersion of the oil.

Native microorganisms have been used in remediation to cleanup environments polluted with oil (Chettri *et al.*, 2016; Tao *et al.*, 2017). Also, a recent discovery revealed a close association of aerobic hydrocarbon-degrading bacteria and phytoplankton. Indeed, a novel PAH-degrading microorganism, *Porticoccus hydrocarbonoclasticus*, was isolated from phytoplankton (Gutierrez *et al.*, 2012a), and members of three major phytoplankton groups (diatoms, dinoflagellates, and coccolithophores) have been found to successfully coexist and even thrive in the presence of obligate and generalist oil-degrading bacteria (Gutierrez *et al.*, 2012a; 2012b; 2014; Green *et al.*, 2015). Some of these interactions are mediated by soluble or volatile metabolites to transmit information between spatially distant species while other interactions involve species in very close proximity (e.g., as a biofilm on the same particle or particles physically associated to one another) (McGenity *et al.*, 2012).

Impacts of Oil on Marine Life

Physical impacts

Oil pollution is a significant source of mortality globally for marine wildlife and the most conspicuous impact of oil spills is the resulting oiled birds, sea mammals, sea turtles, dolphins, corals, sponges, crustaceans, molluscs, and echinoderms. The best documented impacts are large catastrophic spills, which receive ample public attention. However, the more difficult to quantify chronic release of oil caused by oil seeps, leaking shipwrecks, vessels legally and illegally cleaning bilges in marine waters, and other non-point sources can also be a major source of mortality for marine life (Henkel *et al.*, 2014). The following sections provide examples of impacts on some of the marine life affected by oil pollution (see Figure 3).

Marine birds

Large oil spills have devastating effects on birds, for example, it has been estimated that 250,000 birds were killed in the Exxon Valdez oil spill (Piatt and Ford 1996). Although the impact of chronic oiling is much more difficult

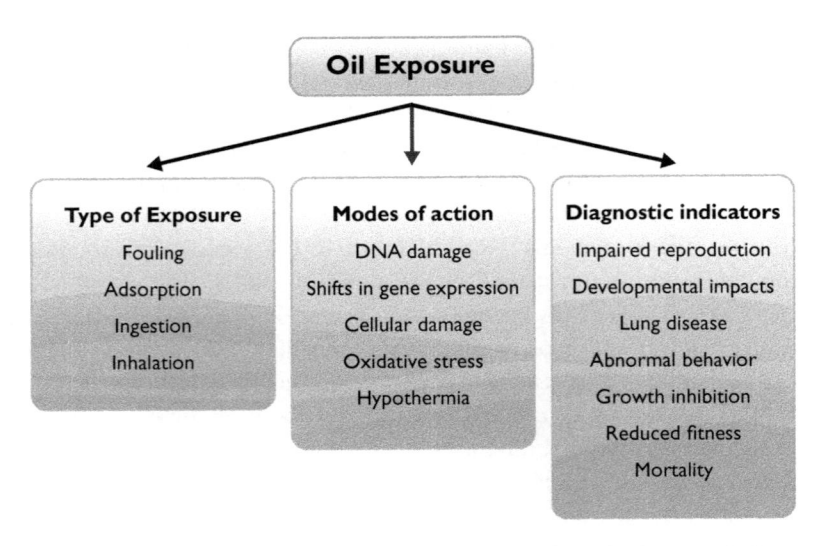

162   *The Future of Marine Life in a Changing Ocean*

```
                        ┌─────────────────┐
                        │  Oil Exposure   │
                        └─────────────────┘
           ┌───────────────────┼───────────────────┐
           ▼                   ▼                   ▼
┌───────────────────┐ ┌───────────────────┐ ┌──────────────────────┐
│ Type of Exposure  │ │  Modes of action  │ │ Diagnostic indicators│
│                   │ │                   │ │                      │
│     Fouling       │ │    DNA damage     │ │ Impaired reproduction│
│    Adsorption     │ │Shifts in gene     │ │ Developmental impacts│
│    Ingestion      │ │    expression     │ │    Lung disease      │
│    Inhalation     │ │ Cellular damage   │ │   Abnormal behavior  │
│                   │ │  Oxidative stress │ │   Growth inhibition  │
│                   │ │    Hypothermia    │ │    Reduced fitness   │
│                   │ │                   │ │      Mortality       │
└───────────────────┘ └───────────────────┘ └──────────────────────┘
```

Figure 3. Types of exposure to oil pollution, impacts on biology, and diagnostic indicators of oil pollution.

to document, it has been estimated that substantial chronic oiling occurs in northern Europe (Lennart Larsen *et al.*, 2007), the southwest Atlantic (Garcia-Borboroglu *et al.*, 2010), and in California (Stenzel *et al.*, 1988; Roletto *et al.*, 2003).

The most vulnerable bird species are those dependent upon feeding in water (e.g., diving birds) compared to semi-aquatic species that can feed ashore. In an oil spill event, oil adheres to the plumage (feathers) reducing the water repellant properties, and causing water to penetrate into the plumage to displace the insulating layer of air. This can expose the animal's sensitive skin to extreme temperatures. In water, oiled feathers can cause heat loss to exceed the bird's heat production capacity, resulting in hypothermia. For example, in order to maintain adequate core body temperature, oiled eiders swimming in cold water have been found to require a 400% increase in their metabolic heat production compared with that of unoiled birds in the same environment (Leighton, 1993).

When birds are covered by oil, they try to get the oil off their feathers by preening, which results in ingestion of oil causing severe damage to internal organs. Also, when the animal is oiled, the focus on preening dominates other natural behaviors, including avoiding predators and feeding, and consequently, oiled birds are also particularly vulnerable to predation, disease, weight loss and dehydration. Many oil-soaked birds lose their buoyancy and drown or beach themselves in their attempt to avoid the cold water.

Sea turtles

Sea turtles have been at risk of exposure in many oil spills, particularly in the Caribbean and Gulf of Mexico. Oil on the sea surface from spills and leaks can get in the eyes, skin, and lungs of sea turtles when they reach the surface to breathe. Not only adults but also eggs on the beach, post-hatchlings and juveniles in the open ocean gyres, and subadults in nearshore habitats are affected by oil pollution, particularly oil-contaminated food. Eggs may become oiled if there is oil in the sand and newly hatched turtles may get contaminated on their way from the beach to the water. Eggs exposed to oil during the last half to last quarter of the incubation period threaten hatchling survival because oil can prevent oxygen from getting through the sand to the eggs and could also change the nest incubation temperature (Lutcavage *et al.*, 1996).

Marine mammals

Like sea turtles, whales, dolphins, manatees, seals, sea otters and other mammals need to come to the surface frequently to breathe air which may expose them to spilled oil on the sea surface. Animals may come in contact with oil through their skin, eyes, nose, mouth, respiratory surfaces, and anal and urogenital orifices. Petroleum hydrocarbons, specifically volatile aromatics and short-chain compounds, are damaging to these and other delicate tissues causing irritation and burns as well as increasing the vulnerability to infections. Additionally, oil can be swallowed or inhaled (volatile compounds), and consumed through prey contaminated with oil. In some marine mammals, fur is an effective thermal barrier because it insulates the animal by trapping air and repelling water. Oil exposure removes natural oils that waterproof the fur and double the rate of heat transfer through fur seal skin (Kooyman *et al.*, 1976, 1977). Some mammals including cetaceans and manatees have no fur and are therefore not susceptible to hypothermia such as those experienced by animals with fur (e.g., sea otters, polar bears, or seals).

The feeding patterns of whales at the sea surface can also be affected by oil. Cetaceans that feed either at the surface or on the ocean floor are more likely to encounter oil than those that generally feed in the water column. Baleen whales appear to be the most vulnerable because of the low population sizes, feeding strategies, and restrictive habitats for feeding and reproduction. For example, fouling of baleen plates used for filtering large volumes of water can decrease their ability to feed (Engelhardt, 1983).

Inhalation of volatile oil compounds may irritate or injure the respiratory tract of mammals which can lead to lung inflammation or pneumonia. Ingestion of petroleum compounds may cause injury to the gastrointestinal tract, which can affect the animals' ability to absorb or digest foods (Engelhardt, 1983). All these impacts can decrease survival and lower the reproductive success of marine mammals.

Marine microorganisms

Microbial degradation plays a major role in the weathering of oil and in the consequent changes in its toxicity that affect phytoplankton (Head et al., 2006). Phytoplankton provide oxygen, dissolved organic matter, and extracellular exudates to bacteria, while bacteria, in turn, release CO_2, exopolysaccharides, vitamins, nutrients, enzymes, and iron to phytoplankton (McGenity et al., 2012) when using hydrocarbons. Therefore, these product exchanges could represent an advantage to phytoplankton cells to survive the toxic effects of oil (Ozhan et al., 2014). Indeed, some studies have shown enhanced degradation of hydrocarbons when bacteria and phytoplankton coexist (e.g., Warshawsky et al., 2007; Abed, 2010).

Oil contamination has a large impact on the abundance and community structure of bacteria. For example, a study conducted during the Deepwater Horizon oil spill showed a shift in the abundance and community composition of indigenous bacteria in beach sands, towards members of the Gammaproteobacteria (*Alcanivorax, Marinobacter*) and Alphaproteobacteria (Rhodobacteraceae), which are important players in oil degradation in the Gulf of Mexico (Kostka et al., 2011). Additionally, the complexity of hydrocarbons present in oil requires resource partitioning by microbial populations, which modify the composition of oil leading to microbial succession (McGenity et al., 2012).

Oil impacts vary among groups of phytoplankton (Gilde and Pinckney 2012; Ladd et al., 2018). Indeed, oil exposure has been shown to negatively impact the growth and motility of some phytoplankton (Østgaard et al., 1984; Garr et al., 2014) although in some cases oil appears not to have an effect (Harrison et al., 1986). The increase of oil in seawater leads to the formation of oil-rich aggregates, involving complex interactions with bacteria, phytoplankton, and zooplankton leading to oil degradation and downward flux to the deep sea floor as marine snow (Beyer et al., 2016). Passow et al. (2012) used roller table experiments to find that oil-rich marine snow is formed through a complex interaction of three mechanisms: (1) production of mucous webs through activities of bacterial oil-degraders associated with floating oil; (2) production of oily particulate matter via coagulation of oil with suspended matter; and (3) coagulation of phytoplankton with oil droplets incorporated into aggregates. This has repercussions in the downward flux of particulate matter to the deep sea and carbon sequestration.

Toxic impacts of oil

Toxic effects of oil are more likely to occur from oil types with high solubility in seawater rather than high-density oil or other highly viscous oil that has low water solubility, and thus its chemical constituents have low bioavailability. The following sections provide some examples of how oil interferes with organism function.

Oil as endocrine disruptor

Organic pollutants including oil can cause endocrine disruption in the aquatic ecosystems (Porte et al., 2006). Animals exposed to these substances experience alterations in hormone synthesis, transport, receptor interaction, metabolism, excretion, or feedback regulation as well as impacts during sex differentiation that may be manifested after sexual maturation (Oberdorster and Cheek, 2001). The term "endocrine disruption" arose following observations of fish displaying intersex behavior in British rivers. Specifically, male fish appeared to have egg-like structures in their testes and higher than expected levels of vitellogenin (Vtg) (a well-known xenoestrogene[1] and a precursor molecule of yolk proteins in oviparous) in their bloodstreams when exposed to effluent of sewage treatment works containing a number of different human-derived steroids such as estradiol and estrogens contained in contraceptive pills (Purdom et al., 1994; Desbrow et al., 1998; Routledge et al., 1998).

---

[1]A type of xenohormone (naturally occurring or artificially made hormone) that resembles estrogen and causes endocrine disruption in animals.

A study exploring the effect of water-soluble fractions of diesel fuel oil and naphthalene in female Atlantic croaker during critical periods of their reproductive life history showed that both pollutants blocked sexual maturation in some fish, impaired ovarian function in others, and most oocytes in exposed fish were undeveloped and malformed at higher concentrations of PAH (Thomas and Budiantara, 1995). Induction of Vtg contaminant in fish has been extensively used as biomarker for contamination in marine animals (Matthiessen and Sumpter, 1998; Ortiz-Zarragoitia and Cajaraville, 2005a). In addition to fish, exposure of blue mussels to North Sea oil, alkylphenols and PAHs, resulted in a significant increase in Vtg-like protein levels in male mussels while female mussels appeared to be unaffected (Ortiz-Zarragoitia and Cajaraville, 2005b). Also, soft shell clams inhabiting PAH-contaminated sites showed anti-estrogenic effects indicated by low Vtg-like protein levels (Gagné et al., 2002). Similarly, female blue mussels exposed to North Sea oil showed significantly lower Vtg-like protein concentrations and gamete development than the controls (Ortiz-Zarragoitia and Cajaraville, 2005b). Both sexes of the mussel *Mytilus galloprovincialis* exposed to three different oils showed impacts on their reproductive health. At high doses, accelerated spawning was observed while at low and medium doses, retarded gonad development and severe alteration of the gonads were observed (Cajaraville et al., 1992).

Genotoxic effects of oil

A genotoxin is a compound that can cause DNA or chromosomal damage. Those compounds such as PAHs that react with DNA are often referred to as genotoxic carcinogens. Many examples of exposure to PAHs have shown damage in embryos, where genotoxic carcinogens have the potential to cause a heritable altered trait (germline mutation). DNA damage in a somatic cell can lead to somatic mutation, which may result in cancer (Phillips and Arlt, 2009). DNA adducts are modifications of DNA that have been exposed to specific carcinogens and thus the level of DNA adducts in cells can serve as a biomarker for a significant exposure to carcinogens. Several biomarkers including enzymatic biomarkers and DNA integrity have been used to assess the extent of genotoxicity in marine organisms (Akcha et al., 2000; Sarker et al., 2018).

PAHs are metabolized by a variety of enzymes, i.e., they convert PAHs to more water-soluble byproducts that are readily excreted from the body. These enzymes include cytochrome P450 (P450 or CYP), epoxide hydrolase, glutathione transferase, UDP-glucuronosyltransferase, sulfotransferase, NAD(P)H quinone oxidoreductase 1, and aldo-keto reductase (Shimada, 2006). A negative side effect, however, is that during catabolism of PAHs, unstable and reactive products attack DNA, and enzymes involved produce carcinogenic metabolites (Shimada, 2006).

Genotoxicity impacts are generally related to the concentration of pollutant, for example, the impact of PAHs on DNA integrity in rock oysters increases with the increasing PAH levels in their tissues (Sarker et al., 2018). Similarly, genotoxicity increased proportionately with exposure to elevated levels of the water accommodated fraction (WAF)[2] of petroleum hydrocarbons in the diatom *Chaetoceros tenuissimus* (Deasi et al., 2010). Amat et al. (2006) showed a positive correlation between the number of DNA adducts and the concentration of PAHs in the surrounding environment. Interestingly, exposure to WAF can also lead to oxidative stress, which is associated with dysfunctional development and reproduction in some marine animals (Han et al., 2014).

What's Next?

To reduce oil pollution, limitations on operational spilling (e.g., when tanks are being cleaned) in addition to international agreements to ensure that regulations are observed are urgently needed. Various technologies of ship and aircraft surveillance including side-looking airborne radar (SLAR), infrared (IR) and ultraviolet (UV) scanners, synthetic aperture radar (SAR) are being used by some countries in oil spill detection systems (Brekke and Solberg, 2005). Although these technological applications can aid in detecting sources of oil pollution, there are still practices that are difficult to survey, including the release of produced water (containing oil and treatment chemicals added during oil production), as well as cleaning operations, small and frequent oil releases from plants, and small operational oil discharges from pipelines during commissioning, maintenance and decommissioning phases that are unaccounted for. Finally, there are examples through history when implementation of some of the regulations that affect the oil industry has been blocked with the arrival of new governments, for example, in the

---

[2]Widely adopted reference fraction for evaluating the toxic effects of oil on marine organisms and for risk assessment of oil spillage events. The method consists of mixing oil and seawater at a given temperature and in the dark (Aurand and Coelho, 2005).

U.S. important laws protecting the environment from industrial malpractice include the Clean Air Act, the Clean Water Act, the Safe Drinking Water Act, the Comprehensive Environmental Response, Compensation, and Liability Act, the Emergency Planning and Community Right-to-Know Act (EPCRA), the Occupational Safety & Health Administration, the Toxic Substances Control Act, the Oil Pollution Act and Spill Prevention Control and Countermeasure Plans. Like other regulations that are intended to regulate the environment, changes in their implementation giving the oil industry (as well as mining, and other major polluting industries) freedom to discharge harmful contaminants can have huge detrimental environmental consequences with extremely lengthy recovery times.

Key Points

- The marine environment is exposed to natural oil released from seeps but humans add a similar amount of crude or refined oil into the oceans from human activities (e.g., oil leaks and spills during the extraction, transportation, refining, storage, and use) which causes serious disruption of the marine environment.
- Oil transformation processes in seawater include evaporation, spreading, dissolution, dispersion, emulsification, photodegradation, biodegradation, sedimentation and oxidation, and some oil from seeps and from spills form tarballs that become stranded on the shoreline.
- Bacteria are the main agents responsible for degrading oil and they appear to degrade the lighter fractions of oil first although many microbes can also break down long chain alkanes and PAHs.
- Life living in the ocean surface is particularly affected by oil pollution. For example, oil adhesion to the plumage of birds can cause hypothermia and oiled fur in seals and other marine mammals removes natural oils that waterproof the fur, and as a result can double the rate of heat transfer through fur skin. Oiled animals can also be exposed to breathing problems and ingestion of oil through prey contaminated with oil or through self-cleaning behavior.
- Toxic impacts are seen through the food chain from microbes to metazoans. Examples of oil toxicity include endocrine disruption (e.g., changes in hormone synthesis, transport, metabolism, feedback regulation, impacts during sex differentiation) and genotoxic effects, which can cause DNA damage in single-cell organisms, as well as embryo and somatic mutation, which may result in cancer.

Questions

1. Give examples of oil-related activities that are significant sources of oil pollution.
2. What are the repercussions of volatile organic compound contamination?
3. Give examples of abiotic and biotic transformation of oil in seawater.
4. What are causes of death in oiled birds?
5. Give examples of how feeding behavior might increase vulnerability and death risk in oil polluted environments in the ocean.
6. How can oil pollution affect the downward flux of particles to the sea floor?
7. Give examples of oil compounds that act as endocrine-disrupting chemicals.
8. What are the differences between the soluble and insoluble fractions of oil in terms of their toxicity for marine organisms?
9. What broad transformations does oil undergo from the natural seep to the sea surface?
10. What are the main biological degraders of oil in the marine environment?

References

Abed RMM (2010) Interaction between cyanobacteria and aerobic heterotrophic bacteria in the degradation of hydrocarbons. *International Biodeterioration and Biodegradation* **64**, 58–64.

Ahmadun F-R, Pendashteh A, Abdullah LC, Biak DRA, Madaeni SS, Abidin ZZ (2009) Review of technologies for oil and gas produced water treatment. *Journal of Hazardous Materials* **170**, 530–551.

Akcha F, Izuel C, Venier P, Budzinski H, Burgeot T, Narbonne J-F (2000) Enzymatic biomarker measurement and study of DNA adduct formation in benzo[a]pyrene-contaminated mussels. *Mytilus galloprovincialis. Aquatic Toxicology* **49**, 269–287.

Al-Lihaibi SS (2003) Photo-oxidation products of petroleum hydrocarbons in the Eastern Red Sea coastal waters. *Environment International* **28**, 573–579.

Amat A, Burgeot T, Castegnaro M, Pfohl-Leszkowicz A (2006) DNA adducts in fish following an oil spill exposure. *Environmental Chemistry Letters* **4**, 93–99.

Andreev PF, Bogomolov AI, Dobryanskii AF, Kartsev AA (1968) Transformation of petroleum in nature. Pergamon, London, p. 482.

Atlas RM (2011) Oil biodegradation and bioremediation: A tale of the two worst spills in U.S. history. *Environmental Science and Technology* **45**, 6709–6715.

ATSDR, Agency for Toxic Substances and Disease Registry (1995) Toxicology information sheets. Available at http://www.atsdr.cdc.gov/N.

Aurand D, Coelho G, Lusby, MD (Eds.) (2005) Cooperative Aquatic Toxicity Testing of Dispersed Oil and the "Chemical Response to Oil Spills: Ecological Effects Research Forum (CROSERF)." Ecosystem Management & Associates, Inc., Technical Report 07-03.

Barron MG, Ka'Aihue L (2001) Potential for photoenhanced toxicity of spilled oil in Prince William Sound and Gulf of Alaska. *Marine Pollution Bulletin* **43**, 86e92.

Beyer J, Trannum HC, Bakke T, Hodson PV, Collier TK (2016) Environmental effects of the deepwater horizon oil spill: A review. *Marine Pollution Bulletin* **110**, 28–51.

Brekke C, Solberg AHS (2005) Oil spill detection by satellite remote sensing *Remote Sensing of Environment* **95**, 1–13.

Cajaraville MP, Marigómez JA, Angulo E (1992) Comparative effects of the water accommodated fraction of three oils on mussels - 1. Survival, growth and gonad development. *Comparative Biochemistry and Physiology — C*, **102**, 103–112.

Chettri B, Mukherjee A, Langpoklakpam JS, Chattopadhyay D, Singh AK (2016) Kinetics of nutrient enhanced crude oil degradation by *Pseudomonas aeruginosa* AKS1 and *Bacillus* sp. AKS2 isolated from Guwahati refinery, India. *Environmental Pollution* **216**, 548–558.

Deasi SR, Verlecar XN, Ansari ZA, Jagtap TG, Sarkar A, Vashistha D, Dalal SG (2010) Evaluation of genotoxic responses of *Chaetoceros tenuissimus* and *Skeletonema costatum* to water accommodated fraction of petroleum hydrocarbons as biomarker of exposure. *Water Research* **44**, 2235–2244.

De Flora S, Bagnasco M, Zanacchi P (1991) Genotoxic, carcinogenic, and teratogenic hazards in the marine environment, with special reference to the Mediterranean Sea. *Mutation Research* **258**, 285–320.

Desbrow C, Routledge EJ, Brighty GC, Sumpter JP, Waldock M (1998) Identification of estrogenic chemicals in STW effluent. 1. Chemical fractionation and in vitro biological screening. *Environmental Science and Tecnhology* **32**, 1549–1558.

Engelhardt FR (1983) Petroleum effects on marine mammals. *Aquatic Toxicology* **4**, 199–217.

Fieser LF, Fieser M (1956) Organic Chemistry, 3rd edition. Boston: DC Heath and Co. Chapter 21.

Furnes G (1994) Discharges of produced water from production platforms in the North Sea. Report R-064641, Norsk Hydro Res. Centre, Bergen, Norway.

Gagné F, Blaise C, Pellerin J, Gauthier-Clerc S (2002) Alterations of the biochemical properties of female gonads and vitellins in the clam *Mya arenaria* at contaminated sites in the Saguenay Fjord. *Marine Environmental Research* **53**, 295–310.

García-Borboroglu P, Dee Boersma P, Ruoppolo V, Pinho-da-Silva-Filho R, Corrado-Adornes A, Conte-Sena D, Velozo R, Myiaji-Kolesnikovas C, Dutra G, Maracini P, Carvalho-do-Nascimento C, Ramos-Júnior V, Barbosa L,

Serra S (2010) Magellanic penguin mortality in 2008 along the SW Atlantic coast. *Marine Pollution Bulletin* **60**, 1652–1657.

Garr AL, Laramore S, Krebs W (2014) Toxic effects of oil and dispersant on marine microalgae. *Bulletin Environmental Contamination and Toxicology* **93**, 654–659.

GESAMP (2007) Estimates of oil entering the marine environment from sea-based activities. Rep. Stud. GESAMP No. 75, 96 p. http://www.gesamp.org/publications/publicationdisplaypages/rs75.

Gilde K, Pinckney JL (2012) Sublethal effects of crude oil on the community structure of estuarine phytoplankton. *Estuaries and Coasts* **35**, 853–861.

Green DH, Echavarri-Bravo V, Brennan D, Hart MC (2015) Bacterial diversity associated with the coccolithophorid algae *Emiliania huxleyi* and *Coccolithus pelagicus* f. braarudii. *BioMed Research International* 2015, doi:10.1155/2015/194540.

Griffin A (1994) MARPOL 73/78 and vessel pollution: A glass half full or half empty? *Indiana Journal of Global Legal Studies* Vol. 1: Iss. 2, Article 10.

Gutierrez T, Nichols PD, Whitman WB, Aitken MD (2012a) *Porticoccus hydrocarbonoclasticus* sp. nov., an aromatic hydrocarbon-degrading bacterium identified in laboratory cultures of marine phytoplankton. *Applied Environmental Microbiology* **78**, 628–637.

Gutierrez, T, Green, DH, Nichols, PD, Whitman, WB, Semple, KT, and Aitken, MD (2012b) *Algiphilus aromaticivorans* gen. nov., sp. nov., an aromatic hydrocarbon degrading bacterium isolated from a culture of the marine dinoflagellate Lingulodinium polyedrum, and proposal of Algiphilaceae fam. nov. *International Journal of Systematic Evolutionary Microbiology* **62**, 2743–2749.

Gutierrez, T, Rhodes, G, Mishamandani, S, Berry, D, Whitman, WB, Nichols, PD, et al. (2014) PAH degradation of phytoplankton-associated Arenibacter and description of Arenibacter algicola sp. nov., an aromatic hydrocarbondegrading bacterium. *Applied Environmental Microbiology* **80**, 618–628.

Hampton S, Kelley PR, Carter HR (2003) Tank vessel operations, seabirds, and chronic oil pollution in California. *Marine Ornithology* **31**, 29–34.

Han J, Won E-J, Hwang D-S, Shin K-H, Leec YS, Leungd KM-Y, Lee S-J, Lee J-S (2014) Crude oil exposure results in oxidative stress-mediated dysfunctional development and reproduction in the copepod *Tigriopus japonicus* and modulates expression of cytochrome P450 (CYP) genes. *Aquatic Toxicology* **152**, 308–317.

Harayama S, Kishira H, Kasai Y, Shutsubo K (1999) Petroleum biodegradation in marine environments. *Journal of Molecular Microbiology and Biotechnology* **1**, 63–70.

Harrison PJ, Cochlan WP, Acreman JC, Parsons TR, Thompson PA, Dovey HM, Xiaolin C (1986) The effects of crude oil and Corexit 9527 on marine phytoplankton in an experimental enclosure. *Marine Environmental Research* **18**, 93–109.

Head IM, Jones DM, Roling WFM (2006) Marine microorganisms make a meal of oil. *Nature Reviews Microbiology* **4**, 173–182.

Henkel LA, Nevins H, Martin M, Sugarman S, Harvey JT, Ziccardi MH (2014) Chronic oiling of marine birds in California by natural petroleum seeps, shipwrecks, and other sources. *Marine Pollution Bulletin* **79**, 155–163.

Jernelöv A (2010) The threats from oil spills: Now, then, and in the future. *AMBIO* **39**, 353–366.

Jerrett M, Burnett RT, Pope III CA, Ito DK, Thurston G, Krewski D, Shi Y, Calle E, Thun M (2009) Long-term ozone exposure and mortality. *New England Journal of Medicine* **360**, 1085–1095.

Judd AG (2003) The global importance and context of methane escape from the seabed. *Geo-Marine Letters* **23**, 147–154.

Kleindienst S, Joye SB (2017) Global aerobic degradation of hydrocarbons in aquatic systems. In: Rojo F (ed.), *Aerobic Utilization of Hydrocarbons, Oils and Lipids. Handbook of Hydrocarbon and Lipid Microbiology.* Springer.

Kooyman GL, Gentry RL, McAllister WB (1976) Physiological impact of oil on pinnipeds. Report N.W. Fisheries Center, *Natural Marine Fisheries Services*, Seattle, WA. p. 23.

Kooyman GL, Davis RW, Castellini MA (1977) Thermal conductance of immersed pinniped and sea otter pelts before and after oiling with Prudhoe Bay crude. In: Wolfe DA (ed.), *Fate and Effects of Petroleum Hydrocarbons in Marine Ecosystems and Organisms.* Pergammon Press, New York, NY, pp. 151–157.

Kostka JE, Prakash O, Overholt WA, Green SJ, Freyer G, Canion A, Delgardio J, Norton N, Hazen TC, Huettel M (2011) Hydrocarbon-degrading bacteria and the bacterial community response in Gulf of Mexico beach sands impacted by the deepwater horizon oil spill. *Applied and Environmental Microbiology* **77**, 7962–7974.

Kvenvolden KA, Cooper CK (2003) Natural seepage of crude oil into the marine environment. *Geo-Marine Letters* **23**, 140–146.

Ladd TM, Bullington JA, Matson PG, Kudela RM, Iglesias-Rodríguez MD (2018) Exposure to oil from the 2015 Refugio spill alters the physiology of a common harmful algal bloom species, Pseudo-nitzschia australis, and the ubiquitous coccolithophore. *Emiliania huxleyi. Marine Ecology Progress Series* **603**, 61–78.

Leahy JG, Colwell RR (1990) Microbial Degradation of hydrocarbons in the environment. *Microbial Reviews* **53**, 305–315.

Leighton FA (1993) The toxicity of petroleum oils to birds. *Environmental Reviews* **1**, 92–103.

Lennart Larsen J, Durinck J, Skov H (2007) Trends in chronic marine oil pollution in Danish waters assessed using 22 years of beached bird surveys. *Marine Pollution Bulletin* **54**, 1333–1340.

Lutcavage ME, Plotkin P, Witherington B, Lutz P (1996) Human impacts on sea turtle survival. In: Lutz P. L., Musick J. A. (eds.), *The Biology of Sea Turtles.* CRC Press, New York, pp. 387–409.

Marshall AG, Rodgers RP (2004) Petroleomics: The next grand challenge for chemical analysis. *Accounts of Chemical Research* **37**, 5359.

Matthiessen P, Sumpter JP (1998) Effects of estrogenic substances in the aquatic environment. In: Braunbeck T, Hinton DE, Streit D (eds.), *Fish Ecotoxicology*, Birkhäuser Verlag, Basel (1998), pp. 319–335.

McDonald BC, de Gouw JA, Gilman JB, Jathar SH, Akherati A, Cappa CD, Jimenez JL, Lee-Taylor J, Hayes PL, McKeen SA, Cui YY, Kim S-W, Gentner DR, Isaacman-VanWertz G, Goldstein AH, Harley RA, Frost GJ, Roberts JM, Ryerson TB, Trainer M (2018) Volatile chemical products emerging as largest petrochemical source of urban organic emissions. *Science* **359**, 760–764.

McGenity TJ, Folwell BD, McKew BA, Sanni GO (2012) Marine crude-oil biodegradation: A central role for interspecies interactions. *Aquatic Biosystems*, **8**, 10 http://www.aquaticbiosystems.org/content/8/1/10.

NRC (National Research Council) (2003) *Oil in the Sea III: Inputs, Fates, and Effects*. Washington, DC: National Academy Press.

Oberdorster E, Cheek AO (2001) Gender benders at the beach: Endocrine disruption in marine and estuarine organisms. *Environmental Toxicology and Chemistry* **20**, 23–36.

Ortiz-Zarragoitia M, Cajaraville MP (2005a) Effects of selected xenoestrogens on liver peroxisomes, vitellogenin levels and spermatogenic cell proliferation in male zebrafish. *Comparative Biochemistry and Physiology — Part C* **141**, 133–144.

Ortiz-Zarragoitia M, Cajaraville MP (2005b) Biomarkers of exposure and reproduction-related effects in mussels exposed to endocrine disruptors. *Archives of Environmental Contamination and Toxicology* **49**, 1–9.

Østgaard K, Eide I, Jensen A (1984) Exposure of phytoplankton to ekofisk crude oil. *Marine Environmental Research* **11**, 183–200.

Owens EH, Taylor E, Humphrey B (2008) The persistence and character of stranded oil on coarse-sediment beaches. *Marine Pollution Bulletin* **56**, 14–26.

Ozhan K, Parsons ML, Bargu S (2014) How were phytoplankton affected by the Deepwater Horizon oil spill? *BioScience* **64**, 829–836.

Pampanin DM, Sydnes MO (2013) Polycyclic aromatic hydrocarbons a constituent of petroleum: Presence and influence in the aquatic environment. In: Vladimir Kutcherov and Anton Kolesnikov (eds.), *Hydrocarbon*, ISBN 978-953-51-0927-3, InTech.

Passow U, Ziervogel K, Asper V, Diercks A (2012) Marine snow formation in the aftermath of the deepwater horizon oil spill in the Gulf of Mexico. *Environmental Research Letters* **7**, doi:10.1088/1748-9326/7/3/035301.

Phillips DH, Arlt VM (2009) Genotoxicity: Damage to DNA and its consequences. *EXS* **99**, 87–110.

Piatt JF, Ford RG (1996) How many seabirds were killed by the Exxon Valdez oil spill? *American Fisheries Society Symposium* **18**, 712–719.

Porte C, Janer G, Lorusso LC, Ortiz-Zarragoitia M, Cajaraville MP, Fossi MC, Canesi L (2006) Endocrine disruptors in marine organisms: Approaches and perspectives. *Comparative Biochemistry and Physiology, Part C* **143**, 303–315.

Pothuluri JV, Cerniglia CE (1994) Microbial metabolism of polycyclic aromatic hydrocarbons. In: Chaudhry GR, (ed.), *Biological Degradation and Bioremediation of Toxic Chemicals*. London: Chapman & Hall. pp. 92–124.

Prince RC (2010) Eukaryotic Hydrocarbon Degraders. In: Timmis KN (ed.), *Handbook of Hydrocarbon and Lipid Microbiology*. Springer, Berlin, Heidelberg.

Purdom CE, Hardiman PA, Bye VJ, Eno NC, Tyler CR, Sumpter JP (1994) Estrogenic effects of effluents from sewage treatment works. *Chemistry and Ecology* **8**, 275–285.

Robotham PWJ, Gill RA (1989) Input, behavior and fates of petroleum hydrocarbons. In: Trett, MW (ed.), *The Fate and Effects of Oil in Freshwater*. Elsevier Applied Science, London, p. 41e80.

Røe Utvik TI (1999) Chemical characterization of produced water from four offshore oil production platforms in the North Sea. *Chemosphere* **39**, 2593–2606.

Roletto J, Mortenson J, Harrald I, Hall J, Grella L (2003) Beached bird surveys and chronic oil pollution in Central California. *Marine Ornithology* **31**, 29–34.

Routledge EJ, Sheahan D, Desbrow C, Brighty GC, Waldock M, Sumpter JP (1998) Identification of estrogenic chemicals in STW effluent 2. In vivo responses in trout and roach. *Environmental Science and Technology* **32**, 1559–1565.

Rumchev K, Spickett J, Bulsara M, Phillips M, Stick S (2004) Association of domestic exposure to volatile organic compounds with asthma in young children. *Thorax* **59**, 746–751.

Sarker S, Vashistha D, Saha Sarker M, Sarkar A (2018) DNA damage in marine rock oyster (*Saccostrea Cucullata*) exposed to environmentally available PAHs and heavy metals along the Arabian Sea coast. *Ecotoxicology and Environmental Safety* **151**, 132–143.

Shimada T (2006) Xenobiotic-metabolizing enzymes involved in activation and detoxification of carcinogenic polycyclic aromatic hydrocarbons. *Drug, Metabolism and Pharmacokinetics* **21**, 257–276.

Solomon GM, Janssen S (2010) Health effects of the Gulf oil spill. *JAMA* **304**, 1118–1119.

Stenzel LE, Page GW, Carter, HR, Ainley DG (1988) Seabird mortality in California as witnessed through 14 years of beached bird carcasses. Stinson Beach, California: Unpublished Report, Point Reyes Bird Observatory.

Sundt RC, Ruus A, Jonsson H, Skarphéðinsdóttir H, Meier S, Grung M, Beyer J, Pampanin DM (2012) Biomarker responses in Atlantic cod (Gadus morhua) exposed to produced water from a North Sea oil field: Laboratory and field assessments, Original Research Article. *Marine Pollution Bulletin* **64**, 144–152.

Tao K, Liu X, Chen X, Hu X, Cao L, Yuan X (2017) Biodegradation of crude oil by a defined co-culture of indigenous bacterial consortium and exogenous *Bacillus subtilis*. *Bioresource Technology* **224**, 327–332.

Thomas P, Budiantara L (1995). Reproductive life history stages sensitive to oil and naphthalene in Atlantic croaker. *Marine Environmental Research* **39**, 147–150.

Tilseth S, Solberg TS, Westrheim K (1984) Sublethal effects of water-soluble fraction of Ekofisk crude oil on the early larval stages of cod. *Marine Environmental Research* **11**, 1e16.

Valentine DL, Mezić I, Maćešić S, Crnjarić-Žic N, Ivić S, Hogan PJ, Fonoberov VA, Loire S (2012) Dynamic autoinoculation and the microbial ecology of a deep water hydrocarbon irruption. *Proceedings of the National Academy of Sciences USA* **109**, 20286–20291.

van Hamme JD, Singh A, Ward OP (2003) Recent advances in petroleum microbiology. *Microbiology and Molecular Biology Reviews* **67**, 503–549.

Vila J, Nieto JM, Mertens J, Springael D, Grifoll M (2010) Microbial community structure of a heavy fuel oil-degrading marine consortium: linking microbial dynamics with polycyclic aromatic hydrocarbon utilization. *FEMS Microbiology Ecology* **73**, 349–362.

Wang Z, Fingas MF (2003) Development of oil hydrocarbon fingerprinting and identification techniques. *Marine Pollution Bulletin* **47**, 423–452.

Warshawsky D, LaDow K, Schneider J (2007) Enhanced degradation of benzo[a]pyrene by *Mycobacterium* sp. in conjunction with green alga. *Chemosphere* **69**, 500–506.

Weschler CJ, Shields HC (1997) Potential reactions among indoor pollutants. *Atmospheric Environment* **31**, 3487–3495.

Widdel F, Knittel K, Galushko A (2010) Anaerobic hydrocarbon-degrading microorganisms: an overview. In: Timmis KN, McGenity T, van der Meer JR, de Lorenzo V (eds.), *Handbook of Hydrocarbon and Lipid Microbiology*. Springer, Berlin Heidelberg, pp 1997–2021.

Wolska L (2008) Determination (monitoring) of PAHs in surface waters: why an operationally defined procedure is needed. *Analytical and Bioanalytical Chemistry* 391, 2647–2652.

Yang C-Y, Chang C-C, Chuang H-Y, Hob C-K, Wub T-N, Chang PY (2004) Increased risk of preterm delivery among people living near the three oil refineries in Taiwan. *Environment International* **30**, 337–342.

Chapter 7

Thoughts on the Effects of Climate Change on Food Security

"It's become more apparent that the real problem with these plastics in the ocean is actually the toxics that are adhering to these plastics, which then have the potential to carry into the food chain because these microplastics are being mistaken for food and many other animals in the ocean."

— Emily Penn, co-founder of Pangaea Explorations and expedition, a pioneering all-women ocean voyage to circumnavigate the globe with the aim of raising awareness of the devastating environmental and health impacts of single-use plastics and toxins in the world's ocean.

Introduction

Food security is defined by the United Nations' Food and Agricultural Organization (FAO) as "a condition when all people, at all times, have physical and economic access to sufficient, safe and nutritious food to meet their dietary needs and food preferences for an active healthy life" (FAO, 1996). Specifically, food security is (i) the availability of sufficient food of appropriate quality, supplied through domestic production or imports; (ii) access by individuals to adequate resources for acquiring appropriate foods for a nutritious diet; (iii) utilization of food through adequate diet, clean water, sanitation, and health care to reach a state of nutritional well-being where all physiological needs are met; and (iv) stability, because to be food secure, a population, household or individual must have access to adequate food at all times (FAO, 1996; see also Wheeler and von Braun, 2013). Marine fisheries make an important contribution to food security, representing a healthy source of proteins, oils, vitamins and minerals (Mohanty et al., 2010). Seafood and fishery products are a valuable source of animal protein — as an example, a portion of 150 g of fish provides at least half of the daily protein requirements for an adult. In 2010, fish accounted for 16.7% of the global population's intake of animal protein and 6.5% of all protein consumed, provided more than 2.9 billion people with almost 20% of their average per capita intake of animal protein, and 4.3 billion people with about 15% of such protein (FAO, 2014). Additionally, marine fisheries provide livelihoods and cultural identities of coastal communities worldwide (Garcia and Rosenberg, 2010). However, as a result of overfishing, and also because of habitat loss and pollution caused by humans, global marine fisheries are underperforming economically (Jackson et al., 2001; Johnson and Welch, 2010; Halpern et al., 2012).

Climate change-driven alterations in the ocean including warming of the sea surface on average by 0.07 °C decade^{-1} during the 20th century (IPCC, 2013), a decrease in seawater pH by 0.1 since the Industrial Revolution (Stocker et al., 2013), and a drop in oxygen levels by more than 2% since 1960 (Schmidtko et al., 2017), have altered the distribution and abundance of marine food sources. Although the magnitude of these changes in the future is hard to predict and depends on whether governments move forward with the business-as-usual (Representative Concentration Pathway or RCP 8·5; IPCC; https://sedac.ciesin.columbia.edu/ddc/ar5_scenario_process/RCPs.html) greenhouse gases emission scenario or a mitigation scenario (Cheung, 2018), we know ecosystems are already changing and will continue to do so in the coming years.

Small-scale fisheries in tropical, less developed, and economically poor regions are especially vulnerable to climate change impacts such as warming (IPCC, 2014). As the demand for fish products is predicted to increase and given the vulnerability of marine ecosystems to climate change, food security is at risk. In addition to overfishing, the oceans are subjected to a variety of stressors including warming, acidification, deoxygenation, sea-level rise and pollution that are causing habitat loss, and altering ecosystem diversity and function which affect food security. For example, ocean warming, acidification, and deoxygenation are known to alter physiological, behavioral, biogeochemical, and ecological processes. A classic example is the shifts observed in the biogeography and relative productivity of anchovy and sardine populations in the Pacific Ocean driven primarily by changes in temperature regimes (Chavez *et al.*, 2003). Also, the expansion of oxygen-minimum zones (OMZs) can reduce biodiversity, shift the composition of food webs, and reduce food security and negatively impact livelihoods. For example, the increase in dead zones in tropical latitudes is causing harmful effects on coral reefs and the community of animals depending on them (Altieri *et al.*, 2017). However, there are some gaps in our understanding of these impacts, for example, the extent to which highly productive nutrient-rich coastal ecosystems and upwelling zones, also associated with OMZs, will continue to support the world's largest fisheries (Breitburg *et al.*, 2018). This chapter does not provide a comprehensive review of all literature but rather discusses the knowns and unknowns of the most relevant climate change impacts on marine food production and distribution, as well as consequences for human food and nutrition security.

Impact of Ocean Acidification on Food Security

The effects of ocean acidification (OA) are and will continue to be widespread across a diversity of marine life although responses will likely vary across spatial scales and taxonomic groups (Kroeker *et al.*, 2010; 2013). Although it has been suggested that organisms that build shells, tests or plates of calcium carbonate ($CaCO_3$), such as corals, mollusks and coccolithophores, are the most vulnerable to ocean acidification (1) (Table 1), there are numerous studies that reveal contrasting outcomes. For example, a 45-year study in the North Atlantic suggested that coccolithophores, a phytoplankton group that produces plates of $CaCO_3$, are increasing in abundance as a result of OA and warming (Rivero Calle *et al.*, 2015). Additionally, pigment analyses at the Bermuda Atlantic Time-Series showed a 38% increase in prymnesiophytes (a group that includes coccolithophores) since the 1990s (Krumhardt *et al.*, 2016). A possible explanation for these and other observations indicating resilience of shelled organisms to OA may be that many calcifiers produce tissues and external organic layers that provide protection to shells and skeletons from dissolution in OA-driven corrosive seawater, enabling calcification to be maintained (Ries *et al.*, 2009; Tunnicliffe *et al.*, 2009). An experiment involving transplanting corals and mollusks along gradients of seawater pH at Mediterranean CO_2 vents revealed that they were able to calcify and grow at even faster than normal rates when exposed to elevated CO_2 levels projected for the next 300 years (Rodolfo-Metalpa *et al.*, 2011). These results give some hope and provide interesting examples of acclimation/adaptation to OA. However,

Table 1. Anticipated and observed impacts of climate-driven phenomena. HABs: Harmful algal blooms; SST: Sea surface temperature.

Climate stressors	Effects	Impacts on food security
Increase in SST	Changes in plankton composition (more frequent HABs?) Spread of disease/parasites to higher latitudes Shifts in biogeography Changes in spp. competition including fish	Effects on abundance and health of grazers Threat of algal biotoxins to human and animal health Changes in food availability Tropicalization of fish populations Changes in dietary diversity and seafood nutrition Fluctuations in food availability and access
Expansion of low-O_2 habitats	Increase in primary productivity Death and displacement of fish and benthic invertebrates Upwelling of low O_2, high CO_2, high-nutrient waters to the sea surface	Selection for organisms that evolved physiological and behavioral adaptations to seasonal low O_2 Decline in survival, growth, and reproduction of animals when O_2 levels are below their tolerance Shifts in fish stock abundance Faster growth and development and smaller fish sizes
Increase in coastal nutrient loads	Excess primary productivity Decrease in O_2 levels and increase in death zones Changes in nutritional composition of phytoplankton	Decrease in diversity and survival of marine fauna Increase in mortality Effects on larval development due to changing nutritional properties of phytoplankton
Increasing discharge of pollutants in coastal waters	Incorporation of pollutants into animal meat Bioaccumulation Biomagnification	Toxicity by ingestion of contaminated seafood by animals and humans Increase in diseases (e.g., immunodeficiency, cancer, decrease in physiological performance)

(*Continued*)

Table 1. (*Continued*)

Climate stressors	Effects	Impacts on food security
Ocean acidification	Decrease in coral reef habitats heavily degraded by warming and ocean acidification	Decrease in productivity in coral reef habitats
	Decreased abundance/change in pteropod biogeography	Lower abundance of fish spp dependent on pteropods

CaCO$_3$-bearing organisms are at risk because the protection of shells or increase in calcification comes at a metabolic cost and, in organisms with weak or no organic protection, the dissolution of exposed CaCO$_3$ will occur as seawater pH falls (e.g., Wood *et al.*, 2008; Jiang *et al.*, 2017).

Mollusks are perhaps among the most vulnerable taxonomic groups experiencing detrimental impacts of OA, particularly during early stages of development (Kurihara, 2008). Indeed, it has been suggested that a decline in their survival and growth could become noticeable by the end of the 21st century under a standard scenario of climate change (IS92a) (Fabry *et al.*, 2008). However, like in many other groups that produce CaCO$_3$, there is great variation even between closely related bivalves (Gazeau *et al.*, 2007), and major differences have been found between studies addressing abrupt impacts versus longer-term effects, particularly depending on whether organisms were acclimated to OA and whether other stressor were present (e.g., nutrient limitation) (Michaelidis *et al.*, 2005; Thomsen *et al.*, 2010).

A consequence of decline in populations of mollusks (e.g., planktonic pteropods and mollusk larvae and corals) is that they are prey to many predatory finfish, potentially resulting in the decrease in abundance of some commercial finfish. Indeed, an organism that has been at the center of attention because it is prey to ecologically and economically important fishes, as well as being central to the diets of birds and whales, is the marine pteropod *Limacina helicina* (Armstrong *et al.*, 2005). Pteropods produce shells of aragonite, a type of CaCO$_3$ that is particularly vulnerable to OA, and observations along the Washington–Oregon–California coast indicated that *L. helicina* shell dissolution relative to preindustrial conditions has doubled, and it is expected to continue to increase by as much as 70% by 2050 along the Northern and Central California current coastal ecosystems (Bednaršek *et al.*, 2014). These ecosystems are particularly susceptible to OA because of seasonal upwelling and the authors argue that while pteropods might still thrive in offshore waters in the future, their populations may be at risk in these coastal region (Bednaršek et al., 2014). However, like in other shell-producing organisms, pteropods produce an organic coating that may protect the shell from dissolution suggesting that pteropods might be able to repair and maintain their shells, although at a likely metabolic cost (Peck et al., 2018). For example, it has been suggested that when the energy budget of pteropods is compromised due to environmental stressors including OA, they can change their swimming behavior, reduce their wing beat frequency, with the potential to increase mortality when they were unable to move vertically and collapsed (Manno et al., 2012). A decrease in populations of pteropods could have major implications on food security as pteropods are major contributors to the diet of zooplankton and commercially important species including North Pacific salmon, mackerel, herring, cod and baleen whales (Willette et al., 2001; Boldt and Haldorson, 2003; Orr et al., 2005).

Impact of Warming on Food Security

Given that most marine animals are cold-blooded, their metabolic rates are strongly impacted by any changes in temperature. The main effect of ocean warming on marine fishes and invertebrates is shifts in the distribution of their populations, generally to deeper waters where cooler refuges exist, and to higher latitudes, causing "tropicalization" (increasing dominance of warm-water species) of catch (Cheung et al., 2012) (Table 1). Indeed, changes in the biogeography of species have already been observed through a global analysis of fisheries catch data using an index called the mean temperature of catch (MTC), calculated from the average preferred temperature of each exploited species weighted by their annual catches. This index appears to have been increasing at a rate of 0.19 °C decade^{-1} between 1970 and 2010 (Cheung et al., 2013; Cheung, 2018), indicating that fisheries have been increasingly catching more warm-water species.

Although there is great variation within taxonomic groups, warming is also known to alter the sensitivity of some organisms to OA. For example, OA and warming can synergistically increase mortality (Prada et al., 2017), and bleaching and productivity loss (Anthony et al., 2008) in corals, but warming can also offset the effects of OA (McCulloch et al., 2012). From a physiological point of view, increasing temperature has a stimulatory effect on metabolism until lethal levels are reached, and all developmental stages are highly sensitive to warming (Byrne, 2011), for example, in invertebrates. Marine fishes are expected to decrease their maximum body size under ocean warming (and also deoxygenation) (Cheung et al., 2013). The generalization of the temperature–size rule (Atkinson, 1994; Ohlberger, 2013) states that elevated temperatures during ontogeny result in faster growth and development but smaller adult body sizes (Atkinson, 1994), and appears to be a broadly observed pattern (Lindmark et al., 2018). Behind this rule is the increased demand for oxygen to support metabolism of fish as temperature increases, but because oxygen supply is limited by the dimensional constraints of the gill (Cheung, 2018), fish usually have smaller maximum body size under warming conditions. Maximum body mass is forecasted to decrease on average across many exploited species by 29% per °C warming (Cheung, 2018), in agreement with observations under warming (Cheung et al., 2013). A shift to fishes with smaller bodies that live in warmer waters is likely to occur, which could impact species interactions and biomas (Audzijonyte et al., 2013).

Warming also appears to be an important factor facilitating the intensification of harmful algal blooms (HABs) (Table 1). For example, observations over three decades of some HAB species (e.g., toxic dinoflagellates) revealed that the duration of bloom seasons significantly increased in many coastal Atlantic regions between 40°N and 60°N, where HABs have begun blooming and expanded in recent decades (Gobler et al., 2017).

Impact of Deoxygenation on Food Security

Ocean warming is causing a decline in the solubility of oxygen and other gases in water masses distant from the atmosphere, in addition to increased stratification, which reduces ocean ventilation, also contributing to hypoxia (Rykaczewski and Dunne, 2010). Additionally, urban runoff and biomass from human waste and agriculture cause coastal eutrophication, increasing primarily nitrogen and phosphorous levels, stimulating the proliferation of algae and ultimately leading to coastal hypoxia, which is increasing and expanding globally (Diaz and Rosenberg, 2008). The rise in productivity in surface waters increases the export of organic matter to deep waters where microbial decomposition consumes oxygen and releases CO_2. In addition to the biogeochemical impacts of deoxygenation (see Chapter 4), low oxygen has negative physiological and behavioral repercussions on marine life (Steckbauer et al., 2011). For example, a decrease in oxygen levels will cause decreases in maximum body weight of fish and species intolerant to hypoxia (e.g. tuna) will experience habitat compression with a possible decline in productivity (Stramma et al., 2010; 2012).

Models based on projections of environmental conditions, habitat types and phytoplankton primary production predict a 30–70% increase in global marine fish catch potential in high-latitude regions and a decline of up to 40% in the tropics (Cheung et al., 2010). Low oxygen is also expected to be threatening to coral ecosystems, which support food security and local economies and indeed it appears that oxygen may increasingly be a key factor in the mortality of corals and the animals associated with coral reefs (Altieri et al., 2017).

Oxygen thresholds for hypoxia vary broadly across taxa. Motile organisms that can escape hypoxic waters tend to display comparably high-oxygen thresholds although fast-moving animals like fish do not necessarily show higher lethal thresholds than those with more restrictive mobility (e.g., crustaceans), suggesting that taxonomic differences can also explain different responses to hypoxia (Vaquer-Sunyer and Duarte, 2008). Organisms moving into adjacent habitats looking for optimal temperatures can be at risk of changes in their predator–prey interactions, and can lead to either increased or decreased prey capture, thereby affecting a predator's functional response (Taylor and Eggleston, 2000).

Extreme hypoxia can result in mass mortality (Table 1) but even mild hypoxia can cause strong shifts in physiological performance, for example, deoxygenation can cause higher blood flow and ventilation or increased concentration of respiratory proteins or binding affinity (Doney et al., 2012; Wu, 2002). These acclimations come at an energetic cost and it is expected that hypoxia can lead to reduced feeding, growth, reproduction, although with highly variable responses among species (Wood et al., 2008; Stumpp et al., 2011).

Impact of Marine Pollution on Food Security

Marine pollution directly affects the quality of seafood (Table 1) and can have lethal and sublethal impacts, and the potential to reduce growth and population size and thus food security. For example, the adverse effects of plastics containing harmful compounds, such as endocrine disruptors (Table 2) and persistent organic pollutants, have been documented at all trophic levels (Bonanno and Orlando-Bonaca, 2018) as plastics and their additives biomagnify[1] through the food chain (Table 1). Perhaps, the main

---

[1] The increasing accumulation of bioactive, often harmful molecules through higher trophic levels in the food chain.

Table 2. Potential effects of developmental exposure to endocrine-disrupting chemicals during the lifetime of humans (from Diamanti-Kandarakis et al., 2009; Bergman et al., 2013).

Prenatal	Childhood	Adolescence	Adulthood	Aged
Genital abnormalities Intrauterine growth restriction	Premature development into puberty Learning differences Behavioral differences Asthma Increased sensitivity to infections Testicular dysgenesis syndrome[a] Obesity			
		Altered puberty Infertility	Premature menopause Breast cancer Genital cancer Parkinson disease Alzheimer disease	

[a]Testicular dysgenesis syndrome includes congenital conditions such as cryptorchidism, hypospadias, oligospermia and testicular cancer.

source of dietary exposure of humans to microplastics is via filter-feeding shellfish, which bioaccumulate[2] plastic particles over time (e.g., Avio et al., 2015), affecting food quality and also potentially human health.

Endocrine-disrupting chemicals including plasticizers (plastic additives), surfactants, and chemicals associated with oil are an increasingly concerning type of pollution (Table 2). Hydraulic fracturing technologies developed over the last 65 years aimed at gaining access to previously inaccessible oil and gas reserves have also raised concerns for the potential contamination of water. This is because approximately 1,000 chemicals that are used throughout the process include many carcinogenic and endocrine-disrupting chemicals (Kassotis et al., 2016) (Table 2). Additionally, although banned in most parts of the world since the end of the 20th century, organic chemicals used heavily in industry and agriculture in the past, such as polychlorinated biphenyls and organochlorine pesticides, still persist in the environment and represent a source of toxins and endocrine-disruptors (Porte et al., 2006). Due to the lipophilic and persistent nature of most endocrine-disrupting chemicals and their metabolites, they can bioaccumulate and biomagnify in marine biota affecting biological processes but particularly reproductive functions (Matthiessen, 2003; Lye, 2000) (Table 2).

Most studies on the effect of endocrine-disrupting chemicals have focused on humans with results pointing towards a link between human exposure to endocrine-disrupting chemicals and male and female reproduction system disorders, infertility, endometriosis, breast cancer, testicular cancer, poor sperm quality and function (Sifakis et al., 2017) (Table 2). As a reference, the US environmental protection agency (EPA) proposes that exposure to the endocrine-disrupting chemical BPA, which is used as a synthetic of plastic and epoxy resins, should not exceed $50\,\mu g\,kg^{-1}\,day^{-1}$, although the actual daily human exposure can be much higher (Taylor et al., 2011; Tzatzarakis et al., 2016; Sifakis et al., 2017). Basheer et al. (2004) reported that the concentration of BPA in supermarket seafood (prawn, crab, blood cockle, white clam, squid and fish) from Singapore was between 13.3 and $213.1\,\mu g\,kg\,wet\,weight^{-1}$ indicating that seafood contaminated with BPA may be the primary route of contamination for humans (Kang et al., 2006).

---

[2]The accumulation of bioactive, often harmful molecules in a living organism.

Oil is also a cause of concern in regard to food security although rather than a widespread problem, oil pollution is concentrated in the marine environment, either in oil seep-rich environments, where oil pollution is chronic, or where oil spills take place as a result of accidents or oil-related activities. Among the families of oil compounds, polycyclic aromatic hydrocarbons (PAHs), which are particularly harmful to marine life, are released during petroleum combustion and spillage and also as a result of other human activities, such as biomass burning, and metal works (Zhang et al., 2008; Baek et al., 1991). Specifically, PAHs have carcinogenic and mutagenic properties (Mastrangelo et al., 1996), and settle in sediments disrupting biota and remaining persistent in the marine environment, particularly in estuarine and coastal environments through the spillage of petroleum, industrial discharges, atmospheric fallout, and urban runoff in both developing and developed countries globally (Neff, 1979; Li et al., 2010).

Wastewaters and landfill leachates are significant sources of anthropogenic contaminants such as pharmaceutical products, pesticides, PAHs, polychlorinated biphenyls (PCBs), and many other toxic and harmful products (Daughton and Ternes, 1999; Brooks et al., 2005). Coastal ecosystems are particularly vulnerable to eutrophication from agriculture and urban activities (Carpenter et al., 1998). For example, phosphorous and nitrogen pollution from fertilizers promotes algal blooms and anoxic/dead zones (Diaz and Rosenberg, 2008), which displaces populations of fish and/or kills fish and other marine animals. In addition to fertilizers and pesticides, livestock antibiotics have negative effects on water quality and pose public health problems for marine life and humans (Kemper, 2008). Indeed, resistance in bacteria from different ecosystems has been increasing at an alarming rate as a result of exposure to antibiotics globally (Alanis, 2005). The marine environment is no exception where antibiotic resistant bacteria, used as bio-indicators of antibiotic pollution, have been reported in colons and gills of fish exposed to sewage effluent (Al-Bahry et al., 2009).

Human Demographics and Food Demand

Half of the human population lives within 60 km of the coast, and this proportion is expected to increase in the coming decades, to reach more than 9 billion by 2050 (UNEP, 2007). Also, given this forecast and growth occurring in the less developed countries, it is expected that 70% of humans will live in urban areas, many in megacities of more than 20 million people (UN-DESA, 2009). This will likely cause an increase in coastal perturbation, and also increase the demand for food. According to the World Health Organization (WHO), the projected increase in humans by 2050 will require an increase in protein of >365 million tonnes between 2010 and 2050 (considering an average human weighting 60 kg and taking into account specific need requirements during growth (Rice and Garcia, 2011). According to the UN-WHO (2002), failure to meet this target and/or irregular distribution of food types and availability would lead to widespread malnutrition and possibly starvation in some areas (Rice and Garcia, 2011).

There is general agreement that marine biodiversity, biogeographic distribution, and abundance (with some species increasing and some decreasing) are changing although the actual global extent of changes in biodiversity remains under debate (Stachowicz et al., 2002; Garcia and Grainger, 2005; Perry et al., 2005; Murawski et al., 2007; Dulvy et al., 2008; Knowlton and Jackson, 2008; Sundby and Nakken, 2008; Arvedlund, 2009). Forecasted changes in fish production (increase at high latitudes and decrease at low/mid latitudes) have considerable regional variations (Barange et al., 2014). Increases and decreases in fish production potential by 2050 are projected to be generally <10% of present yields. However, it has been suggested that in order to meet projected food requirements, the production of fish must increase by approximately 50% from current levels by 2050 (Rice and Garcia, 2011), which would require restoring fisheries.

Different strategies have been proposed including rebuilding stocks to former abundances or selectively intensifying the exploitation of lower trophic levels (Essington et al., 2006), although the latter involves ecological consequences including changes in predator–prey interactions. For example, if substantial amounts of protein were removed from lower trophic levels, the foraging success of many dependent predators would be affected (Daunt et al., 2008). In fisheries models, exploitation rate (μ_t), defined as the percentage of biomass that is removed per year ($\mu_t = C_t/B_t$) where C is the catch (or yield) and B is the available stock biomass in year t (Worm et al., 2009), is broadly used to assess the percentage of a population caught in a period of time. The exploitation rate that provides the maximum sustainable yield (MSY) (μ_{MSY}) for a particular stock is often used for single species and also for multispecies/communities (Worm et al., 2009), where stock biomass is considerably (typically 50–75%) lower than the unfished biomass

(Worm et al., 2009). Total fish catch is predicted to increase toward the multispecies MSY (MMSY) and overfishing leads to excess μ_{MMSY} while rebuilding requires reducing exploitation below μ_{MMSY} (Worm et al., 2009).

Quantity Versus Quality

In addition to changes in fish population abundances in response to long-term warming, with the collapse of some fisheries as a result (e.g., Pershing et al., 2015), an equally important fact is that the quality of marine seafood is declining as seafood has become a vector for the transfer of pollutants and toxic compounds. For example, seafood contaminated with microplastics through bioaccumulation, prey ingestion (Table 1), adherence to the organism's surface or during processing and packaging (Cole et al., 2013; EFSA, 2016) may impact the marine food food web by microplastic biomagnification (Ivar do Sul and Costa, 2014). In European markets, the mussle *Mytilus edulis* and oyster *Crassostrea gigas* contain on average 0.36 and 0.47 microplastic particles per gram of soft tissue respectively at the point of human consumption (Van Cauwenberghe and Janssen, 2014), and in China, a study revealed that commercial bivalves contained 2.1–10.5 microplastic particles per gram of soft tissue (Li et al., 2015). In fish markets from Indonesia and the U.S.A., 28% and 25% of all fish had plastics <4.5 mm in their guts (Rochman et al., 2015). However, despite seafood being a source of microplastic contaminants to the human diet, the levels of microplastics in seafood are neither quantified nor regulated (Ziccardi et al., 2016).

In addition to the increasing health concerns about the safety of marine food consumption, is the vulnerability of many nations that are highly dependent on fisheries. Allison et al. (2009) concluded that the main factors contributing to vulnerability were predicted to be warming, the relative importance of fisheries to the country's economies and diets, and the extent to which society can adapt to potential impacts and opportunities. Among the nations showing a high dependence on fisheries, climate change is predicted to increase productive potential in West Africa and decrease it in South and Southeast Asia (Barange et al., 2014). In many countries and islands where fish can exceed 50% of total animal protein intake (Small Island Developing States, Bangladesh, Cambodia, Ghana, Indonesia, Sierra Leone and Sri Lanka) (FAO, 2016), loss of biodiversity and poor quality of food and decreased availability could cause serious economic problems (Gallo et al., 2018). This is important because more than 53% of the global fish and seafood trade originates in developing countries and ∼260 million people are involved in marine capture fisheries globally (Teh and Sumaila, 2013; Lusher et al., 2017). Interestingly, most countries that are most vulnerable to climate change impacts on their fisheries are also the poorest: they represent only 2.3% of global GDP and 22 of the 33 countries in the most vulnerable quartile are classified as Least Developed Countries (Allison et al., 2009). A decrease in seafood supply in the most vulnerable population — low-income people with fish-dependent diets, may have serious consequences to their nutrition as well as their economy (Kent, 1997).

Questions

1. What are the most vulnerable ecosystems to climate change where food security could be particularly impacted?
2. How is ocean acidification likely to affect shelled organisms?
3. Explain how tropicalization of temperate marine ecosystems can impact trophic interactions.
4. If increasing temperature has a stimulatory effect on metabolism, why do fish tend to decrease their maximum body size under ocean warming?
5. Describe the effects of bioaccumulation and biomagnification of pollution in marine organisms.
6. Give examples of sources of endocrine-disrupting chemical pollution in the marine environment.
7. What are the main factors contributing to vulnerability of nations that are highly dependent on fisheries?
8. How can deoxygenation affect food security?
9. Explain how macronutrient pollution as a result of agricultural practices and other human activities in coastal systems can detrimentally impact food security.
10. What are the negative consequences of exploitation of lower trophic levels as a strategy to rebuild stocks to former abundances?

References

Alanis AJ (2005) Resistance to antibiotics: are we in the post-antibiotic era? *Archives of Medical Research* **36**, 697–705.

Al-Bahry SN, Mahmoud IY, Al-Belushi KIA, Elshafie AE, Al-Harthy A, Bakheit CK (2009) Coastal sewage discharge and its impact on fish with reference to antibiotic resistant enteric bacteria and enteric pathogens as bio-indicators of pollution. *Chemosphere* **77**, 1534–1539.

Allison EH, Perry AL, Badjeck M-C, Adger WN, Brown K, Conway D, Halls AS, Pilling GM, Reynolds JD, Andrew NL, Dulvy NK (2009) Vulnerability of national economies to the impacts of climate change on fisheries. *Fish and Fisheries* **10**, 173–196.

Altieri AH, Harrison SB, Seemann J, Collin R, Diaz RJ, Knowlton N (2017) Tropical dead zones and mass mortalities on coral reefs. *Proceedings of the National Academy of Sciences U.S.A.* **114**, 3660–3665.

Anthony KRN, Kline DI, Diaz-Pulido G, Hoegh-Guldberg O (2008). Ocean acidification causes bleaching and productivity loss in coral reef builders. *Proceedings of the National Academy of Sciences USA* **105**, 17442–17446.

Armstrong JL, Boldt JL, Cross AD, Moss JH, Davis ND, Myers KW, Walker RW, Beauchamp DA, Haldorson LJ (2005) Distribution, size and interannual, seasonal and diel food habits of northern Gulf of Alaska juvenile pink salmon, Oncorhynchus gorbuscha. *Deep Sea Research II* **52**, 247–265.

Arvedlund M (2009) First records of unusual marine fish distributions — can they predict climate changes? *Journal of the Marine Biological Association of the United Kingdom*, **89**, 863–866.

Atkinson D (1994) Temperature and organism size — a biological law for ectotherms? *Advances in Ecological Research* **25**, 1–58.

Audzijonyte A, Kuparinen A, Gorton R, Fulton EA (2013) Ecological consequences of body size decline in harvested fish species: Positive feedback loops in trophic interactions amplify human impact. *Biology Letters* **9**, 20121103.http://dx.doi.org/10.1098/rsbl.2012.1103.

Avio CG, Gorbi S, Milan M, Benedetti M, Fattorini D, d'Errico G, Pauletto M, Bargelloni L, Regoli F (2015) Pollutants bioavailability and toxicological risk from microplastics to marine mussels. *Environmental Pollution* **198**, 211–222.

Baek, SO, Field RA, Goldstone ME, Kirk PW, Lester JN, Perry R (1991) A review of atmospheric polycyclic aromatic hydrocarbons: Sources, fate and behaviour. *Water, Air and Soil Pollution* **60**, 279–300.

Barange M, Merino G, Blanchard JL, Scholtens J, Harle J, Allison EH, Allen JI, Holt J, Jennings S (2014) Impacts of climate change on marine ecosystem production in societies dependent on fisheries. *Nature Climate Change* **4**, 211–216.

Basheer C, Lee HK, Tan KS (2004) Endocrine disrupting alkylphenols and bisphenol A in coastal waters and supermarket seafood from Singapore. *Marine Pollution Bulletin* **48**, 1145–1167.

Bednaršek N, Feely RA, Reum JCP, Peterson B, Menkel J, Alin SR, Hales B (2014) Limacina helicina shell dissolution as an indicator of declining habitat suitability owing to ocean acidification in the California Current Ecosystem. *Proceedings of the Royal Society B* **281**, 20140123.

Bergman Å, Heindel JJ, Jobling S, Kidd KA, Zoeller RT, Jobling SK (Eds). (2013) State of the science of endocrine disrupting chemicals—2012. United Nations Environment Programme and World Health Organization, Geneva. Available: http://apps.who.int/iris/handle/10665/78101.

Boldt JL, Haldorson LJ (2003) Seasonal and geographical variation in juvenile pink salmon diets in the Northern Gulf of Alaska and Prince William Sound. *Transactions of the American Fisheries Society* **132**, 1035–1052.

Bonanno G, Orlando-Bonaca M (2018) Ten inconvenient questions about plastics in the sea. *Environmental Science and Policy* **85**, 146–154.

Breitburg D, Levin LA, Oschlies A, Grégoire M, Chavez FP, Conley DJ, Garçon V, Gilbert D, Gutiérrez D, Isensee K, Jacinto GS, Limburg KE, Montes I, Naqvi SWA, Pitcher GC, Rabalais NN, Roman MR, Rose KA, Seibel BA, Telszewski M, Yasuhara M, Zhang J (2018) Declining oxygen in the global ocean and coastal waters. *Science* **359**, doi: 10.1126/science.aam7240.

Brooks BW, Chambliss CK, Stanley JK, Ramirez A, Banks KE, Johnson RD, Lewis RJ (2005) Determination of select antidepressants in fish from an effluent-dominated stream. *Environmental Toxicology and Chemistry* **24**, 464–469.

Byrne M (2011) Impact of ocean warming and ocean acidification on marine invertebrate life history stages: Vulnerabilities and potential for persistence in a changing ocean. *Oceanography and Marine Biology: An Annual Review* **49**, 1–42.

Carpenter S, Caraco NF, Correll DL, Howarth RW, Sharpley AN, Smith VH (1998) Non point pollution of surface waters with phosphorus and nitrogen. *Ecological Applications* **8**, 559–568.

Chavez FP, Ryan J, Lluch-Cota SE, Ñiquen M (2003) From anchovies to sardines and back: Multidecadal change in the Pacific Ocean. *Science* **299**, 217–221.

Cheung WWL (2018) The future of fishes and fisheries in the changing oceans. *Journal of Fish Biology* **92**, 790–803.

Cheung WWL, Lam VWY, Sarmiento JL, Kearney K, Watson R, Zeller D, Pauly D (2010) Large-scale redistribution of maximum fisheries catch potential in the global ocean under climate change. *Global Change Biology* **16**, 24–35.

Cheung WWL, Meeuwig JJ, Feng M, Harvey E, Lam VWH, Langlois T, Slawinski D, Sun C, Pauly D (2012) Climate-change induced tropicalisation of marine communities in Western Australia. *Marine and Freshwater Research*, **63**, 415–427.

Cheung WWL, Watson R, Pauly D (2013) Signature of ocean warming in global fisheries catch. *Nature* **497**, 365–368.

Cole M, Lindeque P, Fileman E, Halsband C, Goodhead R, Moger J (2013) Microplastic ingestion by zooplankton. *Environmental Science and Technology* **47**, 6646–6655.

Daughton CG, Ternes TA (1999) Pharmaceuticals and personal care products in the environment: Agents of subtle change? *Environmental Health Perspectives* **107**, 907–938.

Daunt F, Wanless S, Greenstreet SPR, Jensen H, Hamer KC, Harris MP (2008) The impact of the sandeel fishery closure on seabird food consumption, distribution, and productivity in the northwestern North Sea. *Canadian Journal of Fisheries and Aquatic Sciences* **65**, 362–381.

Diamanti-Kandarakis E, Bourguignon J-P, Giudice LC, Hauser R, Prins GS, Soto AM, Zoeller RT, Gore AC (2009) Endocrine-disrupting chemicals: An Endocrine Society scientific statement. *Endocrine Reviews* **30**, 293–342.

Diaz RJ, Rosenberg R (2008) Spreading dead zones and consequences for marine ecosystems. *Science* **321**, 926–929.

Doney SC, Ruckelshaus M, Duffy JE, Barry JP, Chan F, English CA, Galindo HM, Grebmeier JM, Hollowed AB, Knowlton N, Polovina J, Rabalais NN, Sydeman WJ, Talley LD (2012) Climate change impacts on marine ecosystems. *Annual Review of Marine Science* **4**, 4.1–4.27.

Dulvy NK, Rogers SI, Jennings S, Stelzenmüller V, Dye SR, Skjoldal HR (2008) Climate change and deepening of the North Sea fish assemblage: A biotic indicator of regional warming. *Journal of Applied Ecology* **45**, 1029–1039.

EFSA (2016) Presence of microplastics and nanoplastics in food, with particular focus on seafood. Panel on contaminants in the food chain. *EFSA Journal* **14** (6), (e04501-n/a). https://doi.org/10.2903/j.efsa.2016.4501.

Essington TE, Beaudreau AH, Weidenmann J (2006) Fishing through marine food webs. *Proceedings of the National Academy of Science of the United States of Sciences of the U.S.A.* **103**, 3171–3175.

Fabry VJ, Seibel BA, Feely RA, Orr JC (2008) Impacts of ocean acidification on marine fauna and ecosystem processes. *ICES Journal of Marine Science* **65**, 414–432.

FAO (1996) Rome declaration on world food security and World Food Summit plan of action, World Food Summit, Rome, 13–17 November 1996.

FAO (2014) The State of World Fisheries and Aquaculture 2014. Rome. p. 223.

FAO (2016) The State of World Fisheries and Aquaculture 2016. Contributing to Food Security and Nutrition for All, Rome. p. 200.

Gallo F, Fossi C, Weber R, Santillo D, Sousa J, Ingram I, Nadal A, Romano D (2018) Marine litter plastics and microplastics and their toxic chemicals components: the need for urgent preventive measures. Environmental Science Europe **30**, https://doi.org/10.1186/s12302-018-0139-z.

Garcia SM, Grainger RJR (2005) Gloom and doom? The future of marine capture fisheries. In Fisheries: A Future? Beddington J. R., Kirkwood G. P. (eds.), *Philosophical Transactions of the Royal Society of London, Series B: Biological Sciences*, **360**, 21–46.

Garcia SM, Rosenberg AA (2010) Food security and marine capture fisheries: characteristics, trends, drivers and future perspectives. *Philosophical Transactions of the Royal Society B* **365**, 2869–2880.

Gazeau F, Quiblier C, Jansen JM, Gattuso J-P, Middelburg JJ, Heip CHR (2007) Impact of elevated CO_2 on shellfish calcification. *Geophysical Research Letters* **34**, L07603. doi: 10.1029/2006GL028554.

Gobler CJ, Doherty OM, Hattenrath-Lehmann TK, Griffith AW, Kanga Y, Litaker RW (2017) Ocean warming since 1982 has expanded the niche of toxic algal blooms in the North Atlantic and North Pacific oceans. *Proceedings of the National Academy of Science*, doi.org/10.1073/pnas.1619575114.

Halpern BS, Longo C, Hardy D, McLeod KL, Samhouri JF, Katona SK, Kleisner K, Lester SE, O'Leary J, Ranelletti M, Rosenberg AA, Scarborough C, Selig ER, Best BD, Brumbaugh DR, Chapin FS, Crowder LB, Daly KL, Doney SC, Elfes C, Fogarty MJ, Gaines SD, Jacobsen KI, Karrer LB, Leslie HM, Neeley E, Pauly D, Polasky S, Ris B, St Martin K, Stone GS, Sumaila UR, Zeller D (2012) An index to assess the health and benefits of the global ocean. *Nature* **488**, 615–620.

Intergovernmental Panel on Climate Change (IPCC) (2013) *Climate Change 2013: The Physical Science Basis. Contribution of Working Group I to the Fifth Assessment Report of the Intergovernmental Panel on Climate Change.* Cambridge Univ. Press, Cambridge, UK, p. 1535.

IPCC (2014) *Climate change 2014: synthesis report.* Contribution of Working Groups I, II and III to the Fifth Assessment Report of the Intergovernmental Panel on Climate Change. Core Writing Team, Pachauri R. K., Meyer, L. A. (eds.), Geneva, Switzerland, IPCC. p. 151.

Ivar do Sul JA, Costa MF (2014) The present and future of microplastic pollution in the marine environment. *Environmental Pollution* **185**, 352–364.

Jackson JBC, Kirby MX, Berger WH, Bjorndal KA, Botsford LW, Bourque BJ, Bradbury RH, Cooke R, Erlandson J, Estes JA, Hughes TP, Kidwell S, Lange CB, Lenihan HS, Pandolfi JM, Peterson CH, Steneck RS, Tegner MJ, Warner RR (2001) Historical overfishing and the recent collapse of coastal ecosystems. *Science* **293**, 629–637.

Jiang L, Zhang F, Guo M-L, Guo Y-J, Zhang Y-Y, Zhou G-W, Cai L, Lian J-S, Qian P-Y, Huang H (2017) Increased temperature mitigates the effects of ocean acidification on the calcification of juvenile Pocillopora damicornis, but at a cost. *Coral Reefs*, doi 10.1007/s00338-017-1634-1.

Johnson JE, Welch DJ (2010) Marine fisheries management in a changing climate: a review of vulnerability and future options. *Reviews in Fisheries Science* **18**, 106–124.

Kang J-H, Kondo F, Katayama Y (2006) Human exposure to bisphenol A. *Toxicology* **226**, 79–89.

Kassotis CD, Tillitt DE, Lin C-H, McElroy JA, Nagel SC (2016) Endocrine-disrupting chemicals and oil and natural gas operations: potential environmental contamination and recommendations to assess complex environmental mixtures. *Environmental Health Perspectives* **124**, 256–264.

Kemper N (2008) Veterinary antibiotics in the aquatic and terrestrial environment. *Ecological Indicators* **8**, 1–13.

Kent G (1997) Fisheries, food security, and the poor. *Food Policy* **22**, 393–404.

Knowlton N, Jackson JCB (2008) Shifting baselines, local impacts, and global change on coral reefs. *PLoS Biology*, **6**, e54.doi:10.1371.

Kroeker KJ, Kordas RL, Crim RN, Singh GG (2010) Meta-analysis reveals negative yet variable effects of ocean acidification on marine organisms. *Ecology Letters* **13**, 1419–1434.

Kroeker KJ, Kordas RL, Crim R, Hendriks IE, Ramajo L, Singh GS, Duarte CM, Gattuso J-P (2013) Impacts of ocean acidification on marine organisms: quantifying sensitivities and interaction with warming. *Global Change Biology* **19**, 1884–1896.

Krumhardt KM, Lovenduski NS, Freeman NM, Bates NR (2016) Apparent increase in coccolithophore abundance in the subtropical North Atlantic from 1990 to 2014. *Biogeosciences* **13**, 1163–1177.

Kurihara H (2008) Effects of CO_2-driven ocean acidification on the early developmental stages of invertebrates. *Marine Ecological Progress Series* **373**, 275–284.

Li Q, Zhang X, Yan C (2010) Polycyclic aromatic hydrocarbon contamination of recent sediments and marine organisms from Xiamen Bay, China. *Archives of Environmental Contamination and Toxicology* **58**, 711–721.

Li J, Yang D, Li L, Jabeen K, Shi H (2015) Microplastics in commercial bivalves from China. *Environmental Pollution* **207**, 190–195. http://dx.doi.org/10.1016/j.envpol.2015.09.018.

Lindmark M, Huss M, Ohlberger J, Gardmark A (2018) Temperature-dependent body size effects determine population responses to climate warming. *Ecology Letters* **21**, 181–189.

Lusher A, Hollman P, Mendoza-Hill J (2017) Microplastics in fisheries and aquaculture. Status of knowledge on their occurrence and implications for aquatic organisms and food safety. *FAO Fisheries and Aquaculture Technical Paper*, p. 615.

Lye CM (2000) Impact of oestrogenic substances from oil production at sea. *Toxicology Letters* **112–113**, 265–272.

Manno C, Morata N, Primicerio R (2012) *Limacina retroversa's* response to combined effects of ocean acidification and sea water freshening. *Estuarine Coastal and Shelf Science* **113**, 163–171.

Mastrangelo G, Fadda E, Marzia V (1996) Polycyclic aromatic hydrocarbons and cancer in man. *Environmental Health Perspectives* **104**, 1166–1170.

Matthiessen P (2003) Endocrine disruption in marine fish. *Pure Applied Chemistry* **75**, 2249–2261.

McCulloch M, Falter J, Trotter J, Montagna P (2012) Coral resilience to ocean acidification and global warming through pH up-regulation. *Nature Climate Change* **2**, 623–627.

Michaelidis B, Ouzounis C, Paleras A, Pörtner HO (2005) Effects of long-term moderate hypercapnia on acid–base balance and growth rate in marine mussels *Mytilus galloprovincialis*. *Marine Ecology Progress Series* **293**, 109–118.

Mohanty BP, Behera BK, Sharma AP (2010) *Nutritional Significance of Small Indigenous Fishes in Human Health*. Bulletin No. 162, Central Inland Fisheries Research Institute, Barrackpore, Kolkata, India.

Murawski S, Methot R, Tromble G (2007) Biodiversity loss in the ocean: How bad is it? *Science* **316**, 1281–1284.

Neff JM (1979) Polycyclic aromatic hydrocarbons in the aquatic environment: sources, fates and biological effects. *Applied Science Publishers*, London, pp. 7–33.

Ohlberger J (2013) Climate warming and ectotherm body size — from individual physiology to community ecology. Functional Ecology 27, 991–1001.

Orr JC, Fabry VJ, Aumont O, Bopp L, Doney SC, Feely RA, Gnanadesikan A, Gruber N, Ishida A, Joos F, Key RM, Lindsay K, Maier-Reimer E, Matear R, Monfray P, Mouchet A, Najjar RG, Plattner G-K, Rodgers KB, Sabine CL, Sarmiento JL, Schlitzer R, Slater RD, Totterdell IJ, Weirig M-F, Yamanaka Y, Yool A (2005) Anthropogenic ocean acidification over the twenty-first century and its impact on calcifying organisms. *Nature* **437**, 681–686.

Peck VL, Oakes RL, Harper EM, Manno C, Tarling GA (2018) Pteropods counter mechanical damage and dissolution through extensive shell repair. *Nature Communications* **9**, doi: 10.1038/s41467-017-02692-w.

Perry AL, Low PJ, Ellis JR, Reynolds JD (2005) Climate change and distribution shifts in marine species. *Science* **308**, 1912–1915.

Pershing AJ, Alexander MA, Hernandez CM, Kerr LA, Le Bris A, Mills KE, Nye JA, Record NR, Scannell HA, Scott JD, Sherwood GD, Thomas AC (2015) Slow adaptation in the face of rapid warming leads to collapse of the Gulf of Maine cod fishery. *Science* **350**, 809–812.

Porte C, Janer G, Lorusso LC, Ortiz-Zarragoitia M, Cajaraville MP, Fossi MC, Canesi L (2006) Endocrine disruptors in marine organisms: Approaches and perspectives. *Comparative Biochemistry and Physiology, Part C* **143**, 303–315.

Prada F, Caroselli E, Mengoli S, Brizi L, Fantazzini P, Capaccioni B, Pasquini L, Fabricius KE, Dubinsky Z, Falini G, Goffredo S (2017) Ocean warming and acidification synergistically increase coral mortality. *Nature Scientific Reports* **7**, doi:10.1038/srep40842.

Rice JC, Garcia SM (2011) Fisheries, food security, climate change, and biodiversity: characteristics of the sector and perspectives on emerging issues. *ICES Journal of Marine Science* **68**, 1343–1353.

Ries J, Cohen A, McCorkle D (2009) Marine calcifiers exhibit mixed responses to CO_2-induced ocean acidification. *Geology* **37**, 1131–1134.

Rivero-Calle S, Gnanadesikan A, Del Castillo CE, Balch WM, Guikema SD (2015) Multidecadal increase in North Atlantic coccolithophores and the potential role of rising CO_2. *Science* **350**, 1533–1537.

Rochman CM, Tahir A, Williams SL, Baxa DV, Lam R, Miller JT, The FC, Werorilangi S, Teh SJ (2015) Anthropogenic debris in seafood: Plastic debris and fibers from textiles in fish and bivalves sold for human consumption. *Scientific Reports* **5**, doi: 10.1038/srep14340.

Rodolfo-Metalpa R, Houlbrèque F, Tambutté E, Boisson F, Baggini C, Patti FP, Jeffree R, Fine M, Foggo A, Gattuso J-P, Hall-Spencer JM (2011) Coral and mollusc resistance to ocean acidification adversely affected by warming. *Nature Climate Change* **1**, 308–312.

Rykaczewski RR, Dunne JP (2010) Enhanced nutrient supply to the California Current Ecosystem with global warming and increased stratification in an earth system model. *Geophysical Research Letters* **37**, doi:10.1029/2010GL045019.

Sifakis S, Androutsopoulos VP, Tsatsakis AM, Spandidos DA (2017) Human exposure to endocrine disrupting chemicals: effects on the male and female reproductive systems. *Environmental Toxicology and Pharmacology* **51**, 56–70.

Schmidtko S, Stramma L, Visbeck M (2017) Decline in global oceanic oxygen content during the past five decades. *Nature* **542**, 335–339.

Stachowicz JJ, Terwin JR, Whitlatch RB, Osman RW (2002) Linking climate change and biological invasions: Ocean warming facilitates non-indigenous species invasions. *Proceedings of the National Academy of Science of the United States of Sciences of the U.S.A* **99**, 15497–15500.

Steckbauer A, Duarte CM, Carstensen J, Vaquer-Sunyer R, Conley DJ (2011) Ecosystem impacts of hypoxia: Thresholds of hypoxia and pathways to recovery. *Environmental Research Letters* **6**, doi: 10.1088/1748-9326/6/2/025003.

Stocker TF, Qin D, Plattner GK, Tignor M, Allen SK, Boschung J, Nauels A, Xia Y, Bex B, Midgley BM (Eds) (2013) *IPCC, 2013: Climate Change 2013: The Physical Science Basis*. Contribution of Working Group I to the Fifth Assessment Report of the Intergovernmental Panel on Climate Change. Cambridge, MA: Cambridge University Press.

Stramma L, Schmidtko S, Levin L, Johnson GC (2010) Ocean oxygen minima expansions and their biological impacts. *Deep-Sea Research Part I. Oceanographic Research Papers* **57**, 587–595.

Stramma L, Prince ED, Schmidtko S, Luo J, Hoolihan JP, Visbeck M, Wallace DWR, Brandt P, Körtzinger A (2012) Expansion of oxygen minimum zones may reduce available habitat for tropical pelagic fishes. *Nature Climate Change*, **2**, 33–37.

Stumpp M, Wren J, Melzner F, Thorndyke MC, Dupont ST (2011) CO_2 induced seawater acidification impacts sea urchin larval development I: Elevated metabolic rates decrease scope for growth and induce developmental delay. *Comparative Biochemistry and Physiology Part A, Molecular and Integrative Physiology* **160**, 331–340.

Sundby S, Nakken O (2008) Spatial shifts in spawning habitats of Arcto-Norwegian cod related to multidecadal climate oscillations and climate change. *ICES Journal of Marine Science* **65**, 953–962.

Taylor DL, Eggleston DB (2000) Effects of hypoxia on an estuarine predator-prey interaction: foraging behavior and mutual interference in the blue crab *Callinectes sapidus* and the infaunal clam prey *Mya arenaria*. *Marine Ecology Progress Series* **196**, 221–237.

Taylor JA, Richter CA, Ruhlen RL, vom Saal FS (2011) Estrogenic environmental chemicals and drugs: mechanisms for effects on the developing male urogenital system. *Journal of Steroid Biochemistry and Molecular Biology* **127**, 83–95.

Teh LCL, Sumaila UR (2013) Contribution of marine fisheries to worldwide employment. *Fish and Fisheries* **14**, 77–88.

Thomsen J, Gutowska MA, Saphörster J, Heinemann A, Fietzke J, Hiebenthal C, Eisenhauer A, Körtzinger A, Wahl M, Melzner F (2010) Calcifying invertebrates succeed in a naturally CO_2 enriched coastal habitat but are threatened by high levels of future acidification. *Biogeosciences* **7**, 3879–3891.

Tunnicliffe V, Davies KTA, Butterfield DA, Embley RW, Rose JM, Chadwick WW (2009) Survival of mussels in extremely acidic waters on a submarine volcano. *Nature Geoscience* **2**, 344–348.

Tzatzarakis MN, Karzi V, Vakonaki E, Goumenou M, Kavvalakis M, Stivaktakis P, Tsitsimpikou C, Tsakiris, I, Rizos AK (2016) Bisphenol A in soft drinks and canned foods and data evaluation. *Food Additives and Contaminants Part B Surveillance* **11**, 1–6.

UN-DESA (2009) Population Division of the Department of Economic and Social Affairs of the United Nations Secretariat, World Population Prospects: The 2008 Revision and World Urbanization Prospects: The 2009 Revision. http://esa.un.org/wup2009/unup/index.asp.

UNEP (2007) GEO 4 Global Environment Outlook. Environment for Development. p. 540.

UN-WHO (2002) Protein and amino acid requirements in human nutrition: report of a joint WHO/FAO/UNU Expert Consultation. *WHO Technical Report Series* **953**, p. 284.

Van Cauwenberghe L, Janssen CR (2014) Microplastics in bivalves cultured for human consumption. *Environmental Pollution* **193**, http://dx.doi.org/10.1016/j.envpol.2014.06.010.

Vaquer-Sunyer R, Duarte CM (2008) Thresholds of hypoxia for marine biodiversity. *Proceedings of the National Academy of Sciences* **105**, 15452–15457.

Wheeler T, von Braun J (2013) Climate change impacts on global food security. *Science* **341**, 508–513.

Willette TM, Cooney RT, Patrick V, Mason DM, Thomas GL, Scheel D (2001) Ecological processes influencing mortality of juvenile pink salmon (*Oncorhynchus gorbuscha*) in Prince William Sound, Alaska. *Fisheries Oceanography* **10**, 14–41.

Wood HL, Spicer JI, Widdicombe S (2008) Ocean acidification may increase calcification rates, but at a cost. *Proceedings of the Royal Society B* **275**, 1767–1773.

Worm B, Hilborn R, Baum JK, Branch TA, Collie JS, Costello C, Fogarty MJ, Fulton EA, Hutchings JA, Jennings S, Jensen OP, Lotze HK, Mace PM, McClanahan TR, Minto C, Palumbi SR, Parma AM, Ricard D, Rosenberg AA, Watson R, Zeller D (2009) Rebuilding global fisheries. *Science* **325**, 578–585.

Wu R (2002) Hypoxia: From molecular responses to ecosystem responses. *Marine Pollution Bulletin* **45**, 35–45.

Zhang Y, Dou H, Chang B, Wei Z, Qiu W, Liu S, Liu W, Tao S (2008) Emission of polycyclic aromatic hydrocarbons from indoor straw burning and emission inventory updating in China. *Annals of the New York Academy of Sciences* **1140**, 218–227.

Ziccardi LM, Edgington A, Hentz K, Kulacki KJ, Kane Driscoll S (2016) Microplastics as vectors for bioaccumulation of hydrophobic organic chemicals in the marine environment: A state-of-the-science review. *Environmental Toxicology and Chemistry* **35**, 1667–1676.

Index

A

acclimation, 40, 62–64, 69, 75, 179, 184
adaptation, 10, 41, 55, 62, 64, 69–70, 75, 102–103, 179
alkalinity, 34, 40
anaerobic ammonium oxidation (anammox), 99, 101
anoxia, 7, 14, 18, 35, 43, 93–94, 98, 100, 117, 131, 151, 160, 185
anthropocene, 31
aquaculture, 5, 15, 118, 122
aragonite, 38, 41, 181
archaea, 11
artificial upwelling, 14
Atlantic Meridional Oceanic Circulation, 60–61

B

bacteria, 7–9, 11, 13, 35, 42, 66, 94, 96, 98, 122–123, 160–161, 164–165, 168, 187
bicarbonate ions (HCO_3^-), 3, 32–40, 44
bioaccumulation, 15, 121, 125, 127, 180, 189
biodiversity, 1, 4, 61, 95, 104, 179, 185, 189
Biological Carbon Pump (BCP), 10–11, 13, 75

biomagnification, 15, 180, 189
bleaching, 10, 54, 67–69, 74, 77, 130, 182
Blue Economy/Oceans Economy, 17
Bohr effect, 102
BPA, 129, 136, 186

C

Ca^{2+}, 36–39
Ca^{2+}-stimulated ATPases, 37
Ca^{2+}/H^+ exchanger, 37
calcium carbonate ($CaCO_3$), 3, 7, 9, 18, 28, 34–36, 38–41, 44, 179, 181
calcification, 7–8, 10, 35–37, 39–40, 66, 69, 130, 179, 181
calcite, 34, 38, 41
calcite, high magnesium, 38, 41
cancer causing/carcinogenic, 3, 5, 15–16, 120, 127, 136, 150, 156, 158, 166–168, 186–187
carbon concentrating mechanism (CCM), 35–36
carbon dioxide (CO_2), 3, 6–7, 9, 10–12, 14, 18, 27–37, 39–44, 54, 60–61, 74, 76, 98, 101–102, 107, 160, 164, 179–180, 183
carbon sequestration, 4, 12–13, 64, 66, 165

carbonate ions (CO_3^{2-}), 32–40, 44
carbonic acid (H_2CO_3), 32–33
carbonic anhydrase, 35, 37
circular economy, 135
Clean Air Act, 156, 168
Clean Water Act, 168
climate change, 2–4, 6–7, 9–10, 15, 17, 29, 43, 56, 60, 64, 68, 70–71, 73, 105–106, 130, 177–179, 181, 189–190
climate stressors, 43, 71, 76, 180–181
CO_2 vents, 41, 179
coccolithophores, 7–9, 32, 36–39, 43, 161, 179
coccoliths, 9, 36
combined stressors, 43
container-deposit legislation, 135
coral bleaching, 10, 67–69, 74, 77, 130
coral reefs, 30, 43, 67–68, 70, 126, 130–131, 179, 181, 184
corals, 9–10, 36–39, 41, 43–44, 54, 65–70, 74, 77–78, 130–131, 161, 179, 181–182, 184
corrosive, 35, 38, 43, 156, 179
crude, 149–152, 155, 157–160, 168
crustaceans, 39, 104, 105, 127, 132, 161, 184
cyanobacteria, 6, 8, 35, 132

D

dead zones, 94, 101, 104, 179, 187
deep sea vents, 41, 44, 179
definition of climate change, 3
degradation, 4, 15, 117, 121, 123, 128, 154, 159–161, 164, 168
degree heating months, 68
denitrification, 10, 98–100
deoxygenation, 69, 73, 75, 93–96, 99, 100–101, 103, 178, 183–184
    impacts on food security, 4, 14, 16, 61, 73, 177–184
diatoms, 8–9, 32, 160–161, 167
dichlorodiphenyltrichloroethane (DDT), 120, 136
dinoflagellates, 8–9, 161, 183

dioxin, 15–16
dispersal of crude oil, 159
dissolved inorganic carbon (DIC), 9–10, 13, 32–34, 36, 39
dissolved organic matter (DOM), 11, 13, 164
diversity, 1, 40, 43, 61, 64–65, 70, 73, 75, 77, 95, 104, 132, 178–180, 188–189
    alpha, 64
    beta, 64
    gamma, 65

E

Earth's energy imbalance (EEI), 54, 57
Earth's radiation budget, 54
echinoderms, 36, 38, 40–41, 44, 161
echinoids, 39
ecosystem function, 56, 65, 70, 131
ectotherms, 69–70
El Niño Southern Oscillation (ENSO), 56, 73, 96
endocrine, 15–16, 117, 121, 128–130, 136, 165, 168, 184–186
endocrine disruptors, 3, 15, 117, 121, 128–131, 136, 165, 168, 184–186
endotherms, 69
entanglement, 116, 126
epigenetic inheritance, 103
eutrophication, 43, 94, 104
exploitation rate (fisheries), 188
export flux, 12

F

fibers, 116, 119–122, 124, 126, 128
fisheries, 5, 15–16, 42, 57, 71, 73–74, 76, 94, 101, 105, 117, 178–179, 182, 188–189
fixed nitrogen, 98
flux of carbon, 67
food insecurity, 15, 56
food security, 4, 14, 16, 61, 73, 177–184, 187
foraminifera, 9, 36, 38

fossil fuels, 1, 10, 14, 28–29, 31, 61, 93, 134
functional group, 7–9

G

genotoxic, 150, 166–168
geoengineering, 12
ghost fishing, 126–127
Great Pacific garbage patch, 117, 136
greenhouse gases, 3, 27, 31, 53–55, 61, 98, 178
groundwater discharge, 2

H

habitat compression, 104–105, 108, 183
habitat degradation, 4
Hawaii ocean time (HOT) series, 30, 97
heat waves, 43, 65, 67, 73, 77
heavy metals, 71, 120, 150
high-nutrient low-chlorophyll (HNLC), 12
hormones, 3, 125, 129, 136, 165, 168
human health, 2–5, 15–18, 31, 127, 156, 158, 177, 186
human population, 1–2, 5, 17, 28, 31, 115, 124, 133, 187
hydrocarbon genotoxicity, 150, 166–168
hydrocarbon-degrading bacteria, 160
hydrocarbons, 9, 15, 71, 149, 150–151, 153–154, 156, 160–161, 163–164, 167, 187
hypoxia, 7, 14, 35, 43, 94–95, 98, 101–105, 107–108, 117, 125, 131, 183–184

I

ice caps, 54, 77
ice melting, 61
immune system/responses, 5, 16, 103
immunotoxicity, 15
implementation of extended producer responsibility (EPR), 134

industrial revolution, 28, 31, 44, 54, 178
invasive species, 125, 131
iron fertilization, 13–14

K

Keeling curve, 29

L

La Niña, Pacific Decadal Oscillation, 56–57
life-cycle stages, 3, 9, 36, 40, 75, 104, 108, 134

M

macroplastics, 116, 121, 126, 127, 132, 136
marine mammals, 69, 72, 105, 125, 161, 163–164, 168
marine birds, 65, 69, 71–72, 77, 105, 124–128, 161–163, 168
marine fisheries, 5, 15, 42, 57, 71, 73–74, 76, 94, 105, 117, 178
marine litter/debris, 116–118, 122, 124–126, 131–136
marine snow, 13, 165
mental health, 2
Mercury, 16, 76
methane, 27, 54–55, 74, 151, 156, 158
microbial carbon pump, 10–12
microfibers, 116, 122, 128
microplastic, 16, 116, 118, 120–125, 127–128, 130
migration (climate), 2
migration (human), 2
migration (poleward), 64, 73–74
misconceptions, 1, 5–6
mollusks, 39, 179, 181

N

N_2 fixation, 8, 10, 66, 99
N_2O, 98–100
nanoplastics, 15, 116, 121, 124, 136
nitrification, 10, 98–99

O

ocean acidification (OA), 3, 6–7, 10, 27–28, 30–44, 63, 66–69, 73–74, 130, 178–179
ocean fertilization, 12–14, 40, 42
ocean gyres, 124–125, 163
ocean heat content, 57, 59, 63, 74
oil platforms, 154
oil pollution act, 168
oil refineries, 158–159
oil seeps, 150–152, 154, 161, 184
oil spills, 14, 122, 150, 152, 154–164, 167–168, 187
oxygen minimum zones (OMZ), 94, 96–101, 105, 107, 179
organic layers, 39, 179
overfishing, 15, 70–71, 105, 178, 189
ozone, 100, 158–159

P

P_{50}, 102, 105–106
Pacific Decadal Oscillation, 57, 96
partial pressure, 31–32, 102
particulate organic carbon (POC), 10–11
pCO_2, 29, 34, 43
penguins, 71–72, 127
persistent organic pollutants (POP), 3, 16, 123, 128, 184
pH, 28–29, 33–35, 40, 98
phenology, 56, 66, 69, 71, 73, 75–77
photo-oxidation, 120, 160–161
photolytic, 123
photosynthesis, 6, 9–10, 13–14, 30, 32, 35, 44, 65–66
phthalates, 119, 129–130, 136
pipeline spills, 155–156
plastic debris "collars", 127
plastic dispersion, 121, 122–123, 125, 131–132, 159–160
plastic fibers, 116, 119, 120–122, 124, 126, 128
plastic hitchhiking/rafting, 117, 124, 131–132
plastic lethal necklace, 127
plastic pollution on corals, 130
plastic sizes, 120–121, 124, 127–128, 132
plastic trophic transfer, 121, 127
plastic weathering, 124, 136
plastic, mechanical degradation, 121, 123
Plasticene, 115
plasticizers, 119, 129, 186
Plastisphere, 131–132
polar bears, 65, 72–74, 163
pollution, impacts on food security, 178, 184
polycarbonate (PC), 119
polychlorinated biphenyls (PCBs), 117, 120, 128, 187
polycyclic aromatic hydrocarbons (PAHs), 15, 150–152, 155, 157, 160–161, 166–167, 187
polyvinylchloride (PVC), 119–121, 127, 129, 136
predator–prey interactions, 103, 105, 107, 184, 188
prey availability, 70–73, 103, 105, 107, 125, 128, 163, 168, 181, 184, 189
primary production, 7, 10–11, 42, 66–67, 101, 107, 184
produced water, 154–155, 167
proton pumping, 39
pteropods, 9, 36, 38, 181

Q

Q_{10}, 66

R

radiation balance, 55
regulation of plastic disposal, 132
reproductive health, 15, 17, 32, 40, 128–130, 166
respiration, 6, 14, 42, 96, 98
responsible industry, 76
RubisCO, 35

S

Safe Drinking Water Act, 168
saturation state of $CaCO_3$ (Ω), 34, 38

sea ice, 57, 71–72, 74, 76, 124
sea level, 3, 54, 60, 74–75, 77
sea surface temperature (SST), 56–58, 62, 68, 71–72, 180
sea turtles, 125–127, 131, 161, 163
silica, 9
silicification, 8
slicks (oil), 154
solubility of gasses, 6, 14, 42, 60–62, 75, 94–96, 98, 183
solubility product for CaCO$_3$ (K^*_{SP}), 38
sperm motility, 40–41
spills (oil), 14, 122, 150, 152, 154–164, 167–168, 187
stratification, 54, 66, 74–75, 77, 94, 183
subtropical gyre, 60, 124–125
sulphur release, 8, 75

T

thermal expansion, 60
thermoplastics, 119, 136
thermoset, 119–120
total alkalinity (TA), 34
toxicity, 125, 127–129, 131, 134, 150–151, 157–158, 160, 164–168, 180, 187, 189
transgenerational plasticity/adaptation, 40, 63, 69, 75, 103

tropicalization of fish populations, 65, 70, 73, 77, 180, 182
tuna, 69–70, 77, 106–107, 183

U

upwelling, 7, 12, 14, 41–44, 75, 97, 100, 179–180, 182

V

vaterite, 38, 41
ventilation, 93–96, 98, 102, 107, 183
visual function, 105
volatile organic compounds (VOCs), 156–158

W

warming, 1–3, 5–7, 10, 14, 27–28, 35, 44–45, 53–57, 59–77, 93, 95–97, 101–102, 104, 107–108, 130, 178–179, 182–183, 189
  hiatuses, 57, 77
water accommodated fraction (WAF), 167

Z

zooplankton, 8, 11, 13, 42, 67, 70–71, 75–76, 116, 125, 164, 182

Lightning Source UK Ltd.
Milton Keynes UK
UKHW051932281219
355999UK00003B/212/P

9 781786 347428